Laboratory Statistics

Laboratory Statistics
Methods in Chemistry and Health Sciences

Second Edition

Anders Kallner
Karolinska University Hospital, Karolinska Institutet,
Stockholm, Sweden

ELSEVIER

Elsevier
Radarweg 29, PO Box 211, 1000 AE Amsterdam, Netherlands
The Boulevard, Langford Lane, Kidlington, Oxford OX5 1GB, United Kingdom
50 Hampshire Street, 5th Floor, Cambridge, MA 02139, United States

British Library Cataloguing-in-Publication Data
A catalogue record for this book is available from the British Library

Library of Congress Cataloging-in-Publication Data
A catalog record for this book is available from the Library of Congress

ISBN: 978-0-12-814348-3

For Information on all Elsevier publications
visit our website at https://www.elsevier.com/books-and-journals

Working together
to grow libraries in
developing countries

www.elsevier.com • www.bookaid.org

Publisher: John Fedor
Acquisition Editor: Kathryn Morrissey
Editorial Project Manager: Katerina Zaliva
Production Project Manager: Vijayaraj Purushothaman
Designer: Matthew Limbert

Typeset by MPS Limited, Chennai, India

Contents

Preface to the Second Edition

The Handbook of Statistical Formulas and Terms used in laboratories was launched with a view to provide a quick source and reminder of correct and useful statistical methods. The idea was to do this without indulging in lengthy theoretical discussions. It might be more important to explain what is required to allow the use of the available procedures and the validity of the outcome. The wisdom of this approach is open for discussion but has been appreciated by many users.

The handbook particularly targets colleagues in analytical laboratories, not necessarily limited to those working in chemistry or health related laboratories. There is an increasing number of colleagues who want to write simple programs as alternatives to commercial statistics packages. Spreadsheet programs have become everybody's toolkit and are popular in solving specific tasks in the laboratory. Such endeavors need interpretable, understandable formulas.

As feedback and comments from many students and colleagues, several sections have been expanded and some new sections added. More formulas and worked examples have been added; many were redesigned and rewritten to make them transparent and useful. This may illustrate a slight change of focus; to provide a tool for creating programs and solutions to everyday statistical procedures used in the analytical laboratory using EXCEL. Accordingly, the subtitle of the handbook has been modified to "Methods in Chemistry and Health Science."

Although "legacy" EXCEL functions usually work fine in new versions of EXCEL, new versions often offer functions with new or improved possibilities to detail calculations. Therefore, all EXCEL function have been updated to 2013 standards. For all practical purposes in the laboratory, the innate statistical functions of EXCEL are satisfactory and can be used in countless combinations to simplify programming.

Standardized and consistent terminology is essential and we rely on the practice developed particularly in ISO, BIPM, and NIST.

The list of literature in the section of further reading has been updated and expanded. This is where the full statistical texts may be found. We have included references to some Internet sources which have been checked.

Anders Kallner
Stockholm 2017

Preface and Introduction to the First Edition

When we have to add, or multiply, even big numbers everything goes almost mechanically. This is a routine work, ..., the true mathematical thinking begins when one has to solve a real problem, that is to say, to identify a mathematical structure that would match the conditions of the problem, to understand principles of its functioning, to grasp connections with other mathematical structures, and to deduce the consequences implied by the logic of the problem. Such manipulations of structures are always immersed into various calculations since calculations form a natural language of mathematical structures.

Michael Heller (2008)

The present "compendium" is for those, who like me are engaged in practical laboratory work and do not have a major in statistical analysis and feel somewhat uncomfortable with the statistical jargon. We frequently face the need to analyze large amounts of data of various origins, collected for various purposes in routine or research work, and have discovered the power of spreadsheet programs in calculations and general data analysis.

Commercial statistical "packages" provide many of the analysis used in the laboratory. By necessity the organization of the data in these packages has to accommodate many different requirements and is perhaps not optimal for a particular practical purpose. Laboratorians often desire to visualize their results graphically and interactively. The availability of spreadsheet programs has eliminated much of the problems and hassle with calculations in statistics, provided simple understandable formulas are available. Indeed simple spreadsheet programming can satisfy most of the necessary calculations and offer simple, efficient, and customized solutions.

The present compendium is not meant to be a "short course" in statistics but a source of a quick reference, repetition, or explanation of formulas and concepts and to encourage development of statistical tools and routines in the research and routine laboratories.

Special attention has been given to expressions that can take different formats but, of course, give the same results. Exposing formulas in different formats may to some extent explain their origin, relation to other procedures, and explain their usage. We have tried to align formulas regarding style and terminology and group them in a logical order. Some formulas in the collection have been edited to facilitate applying in spreadsheet programs.

The selection of formulas in the compendium has developed during several courses in applied statistics for laboratorians and scientists with experimental projects. The number of worked examples is extensive and with tables and figures. Whenever feasible the text makes reference to functions and routines in Microsoft EXCEL.

Formulas have been collected and compared from many different sources, scientific literature, common textbooks and the Internet. It is all out there, cast in different forms and shapes but may be difficult to find. An idea with this compendium is to have most of the statistical procedures used in the laboratory collected in one source.

References to individual sources are not given but a list of contemporary literature is provided.

A threat with preprogrammed routines is that, unless simple rules are violated and thus prevented from use, they will always produce a result. The process of programming and calculating statistical routines has proved to deepen the understanding of the procedures and hopefully diminish erroneous use of established procedures. However, the author takes no responsibility for any erroneous decisions based on calculations using formulas in this compendium.

The comprehensive list of contents and index facilitate the access of the desired concept or procedure.

VOCABULARY AND CONCEPTS IN METROLOGY

Many organizations have invested heavily in formulating internationally acceptable, clear, comprehensive, and understandable definitions of terms in metrology. Superficially this may not seem to have any bearing on statistics. Basically, statistics is one way of formulating and expressing mathematical relationships but we also need to agree on and use definitions of common concepts. The most extensive and internationally recognized list of concepts and their definitions is that created by the joint BIPM, ISO, IEC, IFCC, IUPAC, IUPAP, OIML, and ILAC document *International Vocabulary of Metrology—Basic and General Concepts and Associated Terms (VIM)*, downloadable at http://www.bipm.org/ (accessed 2017-10-02).

The definitions are reproduced *in extenso* from the VIM but some notes have been deleted when pertaining to pure metrological problems.

The author is grateful for the interest and many excellent suggestions from students and other users of previous editions of a compendium that was used in teaching. In particular Prof. Elvar Theodorsson, Department of Clinical Chemistry, University of Linköping, Sweden, has provided healthy and constructive criticism.

Anders Kallner (anders.kallner@ki.se).

Some Notes on Nomenclature

Mathematical formulas may be difficult to decipher but are unambiguous and comprehensive. In this compendium formulas are not so compressed and as a consequence hopefully easier to follow for the nonmathematician. Formulas may be presented in several slightly different forms and formats to facilitate recognition by the reader.

SOME WRITING RULES ARE APPLIED THROUGHOUT THE TEXT

References to functions and formulas in EXCEL are written in capital letters. Functions are defined in English. EXCEL versions in other languages than English usually present the function names on a local language. Several dictionaries are available on the internet, e.g., https://support.office.com/, www.piuha.fi and https://en.excel-translator.de/.

In EXCEL intervals of cells are written *cellA:cellB, A:B* or simply *array.*

The period (full stop) "." is used as decimal sign and a comma "," as the 1,000 separator.

Absolute value: $|a|$, i.e., disregarding the sign.

The number of items or observations is abbreviated n or N.

Fractions (division): $\frac{a}{b}$ or a/b; multiplication $a \times b$.

Square root: $\sqrt{a}$ or explicit $\sqrt[2]{a}$, which allows for higher order roots, e.g., $\sqrt[n]{a}$.

Sum: $a_1 + a_2 + a_3 + \cdots + a_n$ is abbreviated: $\sum_{i=1}^{n} a_i$; In EXCEL *SUM(array)*.

Sum of squares: $(a_1)^2 + (a_2)^2 + \cdots + (a_n)^2 = a_1^2 + a_2^2 + \cdots + a_n^2$ is abbreviated $\sum_{i=1}^{n} a_i^2$; In EXCEL *SUMSQ(a_1:a_n)*.

A squared sum $(a_1 + a_2 + a_3 + \cdots + a_n)^2$ is $\left(\sum_{i=1}^{n} a_i\right)^2$.

Multiplication is abbreviated Π (capital π "pi"). Thus, $\prod_{i=1}^{n} a_i = a_1 \times a_2 \times a_3 \times \cdots \times a_n$. In EXCEL *PRODUCT(array)*.

$>$ is read "larger than," $<$ "smaller than," $\geq$ "larger than or equal to," $\leq$ "smaller than or equal to," $\gg$ "much larger than," and $\ll$ "much smaller than."

Average of a population is μ; of a sample $\bar{x}$ or *xbar*.

Standard deviation of a population is σ; Standard deviation of a sample is $s(x)$, $s(X)$, or simply s if there is no risk for misunderstandings.

Consequently, the standard error of the mean (SEM) is $s(\bar{x})$ or $s(\overline{X})$.

The "robust average," meaning median(X) is written $= x^*$. The asterisk indicates that the value is the result of the "robust algorithm," in this case the median.

Additional abbreviations are explained in the text as appropriate.

Greek letters used for certain purposes (small and capital):

Alpha: α and A; *Beta*: β and B; *Gamma*: γ and Γ; *Delta*: δ and Δ; *Epsilon*: ε and E; *Zeta*: ζ and Z; *Eta*: η and H; *Kappa*: κ and K; *Lambda*: λ and Λ; *My*: μ and M; *Xi*: ξ and Ξ; *Pi*: π and Π; *Rho*: ρ and P; *Sigma*: σ and Σ; *Tau*: τ and T; *Phi*: φ and Φ; and *Chi*: χ and X.

Formulas

1. BASICS

1.1 Logarithms and Exponents

The logarithm of a given number and a given base is the power to which the base must be raised to get the number.

If b is the base and a is a given number, the logarithm is x. This is written $x = {}^{b}\log(a)$. The logarithm of a number can be found in tables but usually provided by calculators.

In many applications the notation *log* refers to 10-logarithms (Briggs), i.e., the base 10, and *ln* refers to e-logarithms or "natural" logarithms with $e = 2.7183$ as the base.

$$\text{If } {}^{b}\log(a) = x, \quad \text{then antilog}(x) = a = b^{x} \tag{1}$$

$$\text{Thus, if } {}^{10}\log(a) = x, \quad \text{then antilog}(x) = a = 10^{x} \tag{2}$$

$$\text{and if } {}^{e}\log(a) = x, \quad \text{then } \ln(a) = x \text{ and antiln}(x) = a = e^{x} \tag{3}$$

$$a \times b = c; \quad \log(a) + \log(b) = \log(c); \quad \frac{a}{b} = c; \quad \log(a) - \log(b) = \log(c) \tag{4}$$

$$\log\left(a^{b}\right) = b \times \log(a) \tag{5}$$

$$\frac{{}^{c}\log(a)}{{}^{c}\log(b)} = {}^{b}\log(a) \tag{6}$$

$$a^{-n} = \frac{1}{a^{n}} \tag{7}$$

$$a^{\frac{1}{n}} = \sqrt[n]{a} = \frac{1}{n} \times \log(a); \tag{8}$$

$$\text{Definition of } e{:}e = \lim_{n \to \infty}\left(1 + \frac{1}{n}\right)^{n} = \sum_{n=1}^{\infty}\frac{1}{n!}$$

EXCEL commands: Natural logarithm: LN(a); antilog: EXP(LN(a)) (cf. Eq. (3)).

Laboratory Statistics. DOI: http://dx.doi.org/10.1016/B978-0-12-814348-3.00001-0

1

10-logarithms (Briggs) $LOG(a)$; antilog: $10^{LOG(a)}$ (cf. Eq. (2)).
Value of $e = e^1$: $EXP(1) = 2.7183$; e^b: $EXP(b)$.

Examples: Let $a = 5$, $b = 10$, $c = 3$ and $n = 2$, then

$$^{10}\log(5) = 0.6990; \quad \text{antilog}(0.6990) = 5 = 10^{0.6990}$$

Since $e = 2.7183$ and $^e\log(5) = \ln(5) = 1.61$; $\text{antiln}(1.61) = 5 = e^{1.61}$
$$= 2.7183^{1.61}$$

$$5 \times 10 = 50; \quad \log(5) + \log(10) = \log(50); \quad 0.6990 + 1 = 1.6990;$$
$$\text{antilog}(1.6990) = 50$$

$$\frac{5}{10} = 0.5; \quad \log(5) - \log(10) = \log(0.5); \quad 0.6990 - 1 = -0.3010;$$

$$\text{antilog}(0.6990 - 1) = 0.5$$

$$\log\left(5^{10}\right) = 10 \times \log(5) = 6.990 \quad \frac{^3\log(5)}{^3\log(10)} = {}^{10}\log(5) = 0.6990$$

$$5^{-2} = \frac{1}{5^2} = \log(1) - 2 \times \log(5) = \text{antilog}(0 - 2 \times 0.699)$$
$$= \text{antilog}(-1.398) = \text{antilog}(0.902 - 2) = 0.04$$

$$8^{\frac{1}{3}} = \sqrt[3]{8} = \frac{1}{3}\log(8) = \text{antilog}\left(\frac{1}{3} \times 0.9031\right) = \text{antilog}(0.3010) = 2$$

Calculation of the "natural" e-logarithms or 10-logarithms, are directly available in spreadsheet programs. If mathematical tables are used, logarithms are conventionally expressed with four decimals to achieve sufficient precision for everyday use and only occupy two pages. Table values can be linearly interpolated if necessary.

1.2   Significance and Rounding of Numerical Values

The precision of a numerical value is laid down in its number of significant digits. This calls for a definition and convention.

1. All digits 1−9 are "significant." The digit zero plays a special role in our numbering system.
2. Zeros, which follow a decimal point, when there are only zeros ahead of the decimal point, are not considered significant digits. However, zeros that follow a decimal point preceded by a nonzero digit are significant. See below (5).

3. The digit zero is significant when it is between two significant digits. Thus, 102 has three significant digits and 1.0204 has five significant digits.
4. A digit "zero" appearing to the left of a digit is only a placeholder if the integer is zero and is therefore not significant. Consequently, 0.0123 has three significant digits and 0.0005 has only one significant digit.
5. A digit "zero" appearing to the right of a digit may or may not be a significant, and must be defined by the user. Note that 1.2300 has five significant digits; 1.20 has three significant digits whereas 100 could have one, two, or three significant digits—the number alone is not enough for a conclusion.

The basic principle of our numbering system is that, without an explicit variability, significant digits imply that the actual value is ± one-half of the value of the last significant digit. For instance, a result of 5 suggests (with no other information) an actual value between 4.5 and 5.4, and 1.23 implies a result between 1.225 and 1.234.

Rounding of a numerical value means replacing it by another value, which is approximately equivalent but has a shorter, simpler, or more explicit representation. More importantly, rounding is an indication of the accuracy of the numerical value. An improper rounding, i.e., using too many or too few digits may be misleading to the reader.

To accomplish a correct rounding, identify the significant digits. Moving from left to right, the first nonzero number is considered the first significant digit. Consider the resolution of the measuring system and the uncertainty of the measurement procedure.

It is important to recognize that constants have an infinite number of significant digits and should never determine the significance of the final result. For instance, if you double something you are multiplying it by 2.000... (infinite); not by 2, which would limit the result to one significant digit (see below). When you use π (pi) or e (the basis of natural logarithms) in calculations, use more significant digits than the significant digits used in other factors or terms in the equation.

In calculations it is important to keep track of the number of significant digits and decimals. Thus, in **multiplication or division**, the *number* of significant digits is important; the number of significant digits in the result shall be the same as the smallest number of significant digits in the factors. For instance, $1.2 \times 8.76543 = 10.51852$ would only allow two significant digits, because one factor has two and the other has five, and the result will thus be allowed to have only two. Thus, the answer is 11 (correctly rounded), which has two significant digits. Note that even though both numbers had at least one decimal place, the result is not allowed to have any; or the result would have too many significant digits. Decimal places as such are not considered in determining the number of significant digits.

In **addition or subtraction** the number of decimal places is important, not the number of significant digits. The number of decimal places in the result shall be the same as the minimum number of decimal places for the terms. For instance, consider $1.2 + 87.6498 = 88.8498$ shall have one decimal place, because one of the terms has only one decimal place. Although the other has six, the result will have the minimum of one. Thus, the answer to the equation would be 88.8 (correctly rounded), which has a single decimal place. The number of *significant* digits in the terms is not considered in the case of addition and subtraction.

Use the following rules to round measurement results, consistent with its significance:

1. When the digit next beyond the one to be retained is less than 5, keep the retained digit unchanged. For example: 2.541 becomes 2.5 if two significant digits were allowed.
2. When the digit next beyond the one to be retained is greater than five, increase the retained digit by one. For example: 2.453 becomes 2.5 if two significant digits were allowed.
3. When the digit next beyond the one to be retained is exactly five, and the retained digit is even, leave it unchanged; conversely if the digit is odd, increase the retained digit by one (even/odd rounding). Thus, 3.450 becomes 3.4 but 3.550 becomes 3.6 when rounded to two significant digits.

   NB: Even/odd rounding of numbers provides a more balanced distribution of results. Rounding by spreadsheet programs may apply—or offer—different rules.
4. When two or more digits are to the right of the last digit to be retained, consider them as a group in rounding decisions. Thus, in 2.4(501), the group (501) is considered to be greater than 5 whereas for 2.5(499), (499) is considered to be less than 5.

A rule often used for logarithms is that the result should have as many decimals as the number of significant digits in the original number. For example log(2) is 0.3 and log(21) is 1.32. This rule is frequently also applied to natural logarithms.

Most important of all the rules is to perform all calculations without rounding and apply the rounding to the final result only.

1.3 Numbers and Rounding in EXCEL

By default EXCEL calculates with 15 significant digits which in most practical situations ensures an acceptable level of correctness in calculations. "Significant digits" are defined as outlined above which particularly has implications on the number of decimals. This can be explored by changing between "number" and "scientific notation." Limiting the number of decimals by the shortcut method does not change the number of significant digits in calculations. It is desirable and possible to limit the number of significant digits both to limit the number of decimals and the number of significant digits in an integer. The most general is *ROUND(number, decimals)*. This function uses the rules above, i.e., numbers ending in less than 5 are rounded "down" the other are rounded up. If necessary the functions *ROUNDUP* or *ROUNDDOWN* can be chosen which round the number away and towards zero, respectively. Particularly when negative numbers are addressed it is important to realize that the rounding will be relative to zero.

A useful feature of these functions is the possibility to round integers, e.g., *ROUND(143; −2)* returns 100 whereas *ROUND(152; −2)* returns 200. Obviously this approach can be used also to limit the number of significant digits in the integer, e.g., *ROUND(152; −1)* from three to two 150. However a tailing zero needs to be defined (see above).

1.4 Derivation—Calculus

The derivative of a function at a given input value describes the best linear approximation of the function near that input value, i.e., the slope of the tangent in that point. Therefore, if the "first derivative" is set to zero and solved, the maximum(s) and/or minimum(s) of the function will be obtained where the tangent is horizontal. In higher dimensions, second, third, etc. derivatives can be calculated. If a second derivative is set to zero the inflexion point of the original function is identified.

The derivative of a function $f(x)$ is written dy/dx, y' or $f'(x)$ and interpreted as the "derivative of y with respect to x."

The partial derivative of a function of several variables is its derivative with respect to one of those variables while the others are held constant.

The partial derivative is written $\partial y/\partial x$.

Examples: The first derivative of a third degree function $y = \frac{1}{3}x^3 - 5x^2 - 11x - 5$ is $dy/dx = y' = f'(x) = x^2 - 10x - 11$ with a maximum and minimum at $x = 5 \pm 6$, i.e., $x_1 = -1$ and $x_2 = +11$, respectively.

The second derivative is: $d^2y/dx^2 = y'' = f''(x) = 2x - 10$ and the inflexion point of the original function is $x = 5$.

Draw the three functions and confirm the maximum, minimum, and inflexion points!

If $y = n \times x^k$, the first derivative will, in general terms, be the exponent (k) of the independent variable times its factor (n). The exponent of the independent variable (x) is reduced by one and any constant becomes zero. Thus:

$$\frac{dy}{dx} = n \times k \times x^{(k-1)} \qquad (9)$$

For a detailed discussion of derivative rules, derivatives, and partial derivatives the reader is referred to special literature.

1.5 Trigonometry

In a right-angled triangle, i.e., a triangle with one angle equal to 90 degrees, i.e., one side perpendicular to another side, the sides surrounding the right angle are called *cathetus* (*a* and *b* in Fig. 1A, often known as *leg*) and the opposite side the hypotenuse (*c*). The relation between these sides is expressed by the Pythagoras' theorem:

$$a^2 + b^2 = c^2.$$

The proportions or "image" of any triangle are determined by the angles ($A = BAC$, $B = ABC$, and $C = ACB$). In a right-angled triangle, the sizes of the angles are defined by the trigonometric functions referring to a right-angel triangle (Fig. 1A):

$$\sin A = \frac{a}{c}; \quad \sin B = \frac{b}{c}; \quad \cos A = \frac{b}{c}; \quad \cos B = \frac{a}{c}$$

$$\tan A = \frac{a}{b}; \quad \tan B = \frac{b}{a}; \quad \cot A = \frac{b}{a}; \quad \cot B = \frac{a}{b}$$

The relations between the trigonometric quantities, e.g., angles and other elements of triangles, are known as the Law of sines, Law of cosines, and Law of tangents. The further discussion of trigonometry is outside the scope of this text.

Provided the angle is known and expressed in *radians* EXCEL provides numerical values of these quantities *SIN(A)*, *COS(A)*, and *TAN(A)*. The cotangent for an angle is the inverse of its tangent and is not available as a separate function in EXCEL.

The *unit circle* is a circle with a radius of one. *Radian* is defined as the angle *AOB* (θ) in the unit circle (Fig. 1B), which corresponds to the arc *AB* with a length equal to the radius $OB = OA$. Since the circumference is $2 \times$ radius $\times$ pi(π) and corresponding to 360 degrees, an angle corresponding

(A)

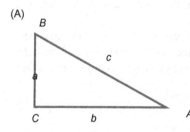

(B)

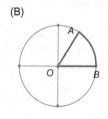

FIGURE 1 (A) Right-angled triangle. (B) The unit circle.

to one radian will be $360/(2 \times 1 \times \pi)$, i.e., 57.3 degrees. If the unit circle is drawn in a coordinate system it is focused at the origin (0;0), therefore sin $(\theta) = x$ and $\cos(\theta) = y$, in all quadrants, with the appropriate signs.

EXCEL provides conversions between degrees and radians: *RADIANS* (angle in degrees) and *DEGREES* (angle in radians), respectively. Therefore, to express the sine of 30 degrees, the function would be *SIN(RADIANS (30)) = SIN(0.52) = 0.5*. The reverse of the trigonometric functions are *arcsine*, *arccosine*, and *arctangent*, respectively. In EXCEL the functions are *ASIN(A)*, *ACOS(A)*, and *ATAN(A)*. Thus, to convert a sine of 0.5 to degrees the function would be *DEGREES(ASIN(0.5))*, yielding 30 degrees.

1.6 Scales—Types of Data

Data can be expressed as four types according to the scales used: nominal-, ordinal-, interval-, and ratio scale.

Data on a nominal scale may be numbers or any other information that describes a property. There is no size relation between the entities.

Data expressed on an ordinal scale are of different sizes and can thus be ordered or ranked, e.g., "bad," "good," "superior," or 1, 2, 3. The scale is arbitrary, i.e., the intervals between numbers may be unequal or even undefined. Data expressed on an ordinal scale can be measured and are thus quantities. Not all statistical procedures can be applied to ordinal data, essentially ordinal data can only be ranked, i.e., ordered.

Data with equal intervals between numbers are of two kinds, and can be expressed on interval- or ratio scales. The ratio scale is characterized by— apart from equally sized units—a natural zero whereas the interval scale has an arbitrarily defined zero. Interval data are characterized by a constant difference between results whereas rational data display constant ratios between data. A commonly cited quantity expressed on an interval scale is temperature expressed as degree Celsius (°C) (previously centigrade) or Fahrenheit (°F) whereas temperature expressed in kelvin (K) are expressed on the ratio scale. Consequently, 40 K is twice as much as 20 K, whereas 40°C is not twice as much as 20°C. However, there are as many degrees between 40 and 20°C as between 20 and 0°C. 1°C exactly equals 1 K, by definition. Note that metrologically the "kind-of-quantity" is temperature which can be expressed in different units, kelvin for the Kelvin scale and degree for the Celsius and Fahrenheit scales.

2. DISTRIBUTIONS OF DATA

It is not often possible to study an entire population; rather we make a selection and study a sample of the population. We can select several different samples from a population. It is not likely that they are identical. If each member of the population has an equal chance of being selected, the selection is called unbiased and representative. It is unlikely that a selection will be absolutely unbiased and, as we will see, there are rules, strategies, and procedures to reduce the bias.

In statistical literature a "parameter" is observed data whereas "statistics" are calculated. Thus, population and quantities of a population are parameters whereas properties and quantities of a sample are statistics and used to estimate parameters of populations.

It is important to understand that it is not possible to do the process backwards, i.e., to calculate the quantity of an individual from the population information without introducing a term of uncertainty.

To illustrate a population the individual data are distributed along a value axis. Although there may not be two identical numbers, observations may be allocated to defined groups or intervals. These intervals may contain different amounts (often referred to as frequencies) of data; a "frequency distribution" will be obtained. The frequency distribution may take any shape (normal, rectangular, logarithmic, etc.) and there are certain distributions which are more interesting than others because they show a regularity which allows them to be treated mathematically. A distribution will be characterized by at least a location parameter (central tendency) and a measure of the dispersion of the data, e.g., variance interquartile interval.

2.1 Histogram

A histogram is a graphical display of data with no conditions specified regarding their distribution. To create a histogram, the data are sorted according to size and allocated to, or grouped in, suitable intervals. The frequency or number in each group, or "bin," is displayed along the vertical (value) axis. The resolution, i.e., the details of the distribution, depends largely on the size and number of bins. Usually the bin sizes are made equal but that is not always the case, they may be adjusted to emphasize certain aspects of the distribution.

Designing a histogram manually is easy but tedious and EXCEL offers two different possibilities. The simpler is to activate the "Data analysis" function which is an add-in to EXCEL and found under the Data tab. This is straightforward but has the disadvantage of not allowing modifications interactively.

A fully flexible procedure can be coded by the "FREQUENCY" function. In short, define the desired bins (B), mark a set of cells, one cell larger than the number of bins, and write $= FREQUENCY(A1:AN_1, B1:BN_2)$ in the first cell. Then press Control + Shift + Enter to create an array. The marked cells become filled with the formula and subsequently with the number of items (frequency) in each bin. The contents of the dataset $(A1:AN_1)$ and the design and contents of the bins $(B1:BN_2)$ can be changed. To modify the array formula, Press F2, make the changes in the first cell and the change by combining Control + Shift + Enter. Any changes will be immediately reflected in the frequency. A bar graph of the histogram can then be displayed.

The frequency assigned to a bin is the sum of observations in the bin below the upper limit of the bin. If used for drawing a histogram a better representation is achieved if the bin in the graph is defined by the average of two adjacent bin limits on the value axis. If this correction is not made, the histogram will appear shifted one half bin-width to the right. This can be noted if an experimental (simulated) normal distribution is compared to a Gauss curve calculated using formula (36).

2.2 The Central Tendency and Dispersion

In the following formulas, x_i represents the result of an observation and n the number of observations.

Thus x_n is the nth observation of the series.

$$\text{Mean, arithmetic } \bar{x} = \frac{\sum_{i=1}^{i=n} x_i}{n}; \bar{x} = \frac{x_1 + x_2 + \cdots + x_n}{n} \quad (10)$$

The mean, or average, answers the question: *If all quantities had the same value, what would that be to achieve the same total sum?*

EXCEL offers the function $AVERAGE(x_1, x_2, \ldots, x_n)$ which directly returns the arithmetic mean.

The average is one of the two variables in the Gaussian distribution function and defines the location of the distribution. It can be thought of as the "center of gravity" of a distribution but it is not necessary a real number in the sense that it physically appears in the studied dataset. It is often used colloquially as a "summary" of a dataset and representing the population/sample but not necessarily used as stringent as the definition in statistics. It can always be calculated but if applied to data which are not normally distributed the result has lost these properties and may be the cause of misunderstandings. A single high or low value (outlier) may have a profound effect on the average.

Example: The average of observations $x_1 = 3$, $x_2 = 4$, $x_3 = 7$, and $x_4 = 10$ is

$$\frac{\sum_{i=1}^{4} x_i}{n} = \bar{x} = \frac{3 + 4 + 7 + 10}{4} = \frac{24}{4} = AVERAGE(3, 4, 7, 10) = 6.$$

The sum of $6 + 6 + 6 + 6 = 24$.

$$\frac{\sum_{i=1}^{4} x_i}{n} = \bar{x} = \frac{3 + 4 + 7 + 98}{4} = \frac{112}{4} = 28.$$

The sum of $28 + 28 + 28 + 28 = 112$ but 28 is not representing the dataset. The average is influenced by the single high value.

The abbreviation $\bar{x}$ is known as "sample average" whereas the "population average" is abbreviated μ.

In the present text we prefer using "average" rather than "mean" except in recognized phrases, e.g., arithmetic mean. Average, not mean, is used in VIM (International vocabulary of metrology-Basic and general concepts and associated terms).

Mean, geometric:
$$\bar{x}_G = \sqrt[n]{(x_1) \times (x_2) \times \cdots \times (x_n)} = \sqrt[n]{\prod_{i=1}^{n}(x_i)}$$

$$\ln(\bar{x}_G) = \frac{1}{n} \times (\ln(x_1) + \ln(x_2) + \cdots + \ln(x_n))$$

(11)

The geometric mean answers the question: *If all quantities had the same value, what would that be to achieve the same product?* In a perfect normal distribution of positive numbers the arithmetic and geometric means are equal.

In mathematical terms the geometric mean is explained as: "The nth root of the product of n numbers."

EXCEL offers the function *GEOMEAN($x_1, x_2, \ldots, x_n$)* which directly returns the geometric mean.

Example: The geometric mean of the above example is

$$\bar{x}_G = \sqrt[4]{(3) \times (4) \times (7) \times (10)} = \sqrt[4]{840} = 5.38.$$

This expression may be more conveniently calculated using logarithms:

$$\ln(\bar{x}_G) = \frac{1}{4} \times (\ln(3) + \ln(4) + \ln(7) + \ln(10)) = \frac{1}{4} \times (1.10 + 1.39 + 1.94 + 2.30) = 1.68,$$

anti In(1.68) = exp(1.68) = 5.38 or *GEOMEAN(3,4,7,10)* = 5.38.
The product $5.38 \times 5.38 \times 5.38 \times 5.38 = 5.38^4 = 840$.

The geometric mean of two numbers is the square root of their product.

A consequence is that the antilogarithm of the arithmetic mean of a logarithmic distribution will be the geometric mean of the underlying distribution. The geometrical mean therefore plays a role when dealing with skew distributions (e.g., in serology) after logarithmic transformation. Another example is describing the size distribution of blood corpuscles.

The geometrical mean is often used in calculations of interests and capital growth but has also several geometrical meanings. If, for instance, a tank measuring $5\ m \times 12\ m \times 2\ m$, thus containing $120\ m^3$ should be exchanged for a cubical structure, the length of the sides (a) would be: *GEOMEAN(5, 12, 2)* $= \sqrt[3]{120} = 120^{1/3} = 120^\wedge(1/3) = 4.93\ m$.

NB: A meaningful geometrical mean can only be calculated from positive numbers. Logarithms cannot be calculated from negative numbers, or from 0. To satisfy the description, of the geometrical mean which includes negative values it can only be calculated from an odd number of observations, resulting in an "odd" root e.g., $\sqrt[3]{-27} = -3$; $(-3) \times (-3) \times (-3) = -27$. It may be difficult to understand and visualize the meaning of a geometrical mean that includes negative values.

$$\text{Mean, harmonic: } \bar{x}_h = \frac{n}{\sum_{i=1}^{i=n}\left(\dfrac{1}{x_i}\right)} \tag{12}$$

To help memorize the calculation it is often written inversely: $\frac{1}{\bar{x}_h} = \frac{1}{n} \times \sum_{i=1}^{n}\left(\frac{1}{x_i}\right)$.

Example 1: The harmonic mean of 10 and 20:
$$\bar{x}_h = \frac{2}{\left(\dfrac{1}{10} + \dfrac{1}{20}\right)} = \frac{2}{\dfrac{2+1}{20}} = \frac{40}{3} = 13.33$$

Example 2: A car travels 100 km at 80 km/h, 100 km at 120 km/h, and 100 km at 150 km/h. Calculate the average speed (V)!
$D = V \times T = 100 + 100 + 100 = 300$ km. $T = (100/80 + 100/120 + 100/150) = 330/120 = 2.75$ h. The average speed (V) is 109 km/h.
Or

$$\frac{1}{\bar{x}_h} = \frac{1}{3} \times \left(\frac{1}{80} + \frac{1}{120} + \frac{1}{150}\right) = 0.009167; \quad \bar{x}_h = 109 \text{ km/h}$$

The harmonic mean will reduce the influence of extreme, large values but increase that of small values. The harmonic mean is equivalent to the inverse mean of the reciprocals of the values and is the best estimate of the average of rates or velocities.

The geometric mean is always the smallest of the three arithmetic, harmonic, and geometric means, the arithmetic always the largest and the harmonic in between.

$$\text{Mean, weighted: } \bar{x}_w = \frac{\sum_{i=1}^{i=k}(\bar{x}_i \times n_i)}{N} = \frac{\sum_{i=1}^{i=N} x_i}{N} \tag{13}$$

where $\bar{x}_i$ is the value of the quantity in each of k bins, n_i is the number of observations in each of the corresponding bins and N is the total number of observations, i.e., $N = \sum_{i=1}^{i=k} n_i$.

Example: A dataset consists of five groups with 3, 3, 6, 2, and 7 items, respectively, $n = 21$: The averages of the corresponding groups were 5, 9, 3, 8, and 4.

$$\bar{x}_w = \frac{3 \times 5 + 3 \times 9 + 6 \times 3 + 2 \times 8 + 7 \times 4}{3 + 3 + 6 + 2 + 7} = \frac{104}{21} = 4.952 \text{ (correctly rounded: 5.0)}$$

$$\text{Standard deviation, population: } \sigma = s(X)_p = \sqrt{\frac{\sum_{i=1}^{i=n} (x_i - \mu)^2}{n}} \qquad (14)$$

The width of a normal distribution is described by the standard deviation $s(X)_p$. This is also known as the biased population standard deviation.

In the graphical representation of the normal distribution—the familiar bell-shaped curve—it is the distance between the peak (average) and the first inflexion point where an increasing negative slope (above the average) changes and the negativity begins decreasing. Since the function is symmetrical around the average the distance to the inflexion point on the other side of the average is the same.

NB: The standard deviation is the square root of a number and is always the positive root.

$$\text{Standard deviation, sample: } s(X)_s = \sqrt{\frac{\sum_{i=1}^{i=n} (x_i - \bar{x})^2}{n - 1}} \qquad (15)$$

The sample standard deviation approaches the population standard deviation, s_p, if the number of observations is large and equals the root mean square (Eq. 24) if $\bar{x}$ is zero [0], e.g., in symmetrical wave functions.

In EXCEL the function is *STDEV.P(array)* and *STDEV.S(array)* for the standard deviation of a population and sample, respectively (Fig. 2).

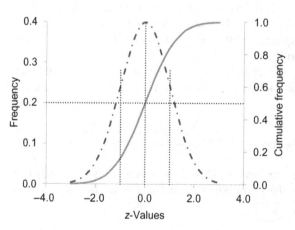

FIGURE 2 Frequency (hatched) and cumulative frequency (solid). z-Values are the number of standard deviations. The vertical dotted line represents the average and $\pm s(x) = \pm 1z$. The dotted horizontal line crosses at the median/average (cumulative frequency 0.5).

The standard deviation represents an interval on the X-axis and is expressed in the same units as the quantity value. Due to the shape of the normal distribution curve the "first standard deviation" counted from the average will cover about 34.1 % of the area under the curve (AUC) (see Table 2), the next only about 13.6, i.e., together about 47.7 %. Therefore, the interval from $-2s$ to $+2s$ includes about 95.4 %, leaving about 2.5 % below $-2s$ and 2.5 % above $+2s$ (Table 1). The "third *standard deviation*" will cover only about 2.1 % of AUC. The z-values (Eq. 59) in Table 1 were obtained by EXCEL function *NORM.S.INV(probability)*. The cumulative area under the curve and frequency (the y-value corresponding to an x-value) are obtained by *NORM.S.DIST(z,true)* and *NORM.S.DIST(z,false)*, respectively.

NB: The definitions of the standard deviation in formulas (14) and (15) above are only applicable to normal distributions, a standard deviation can also be calculated for rectangular and triangular distribution and estimated from other distributions and will then have different properties.

Homoscedasticity—heteroscedasticity. Homoscedasticity is when the variance and thus the standard deviation is constant within a measuring interval. The opposite is heteroscedasticity, which is understood as if the relative variance/relative standard deviation is constant in the measuring interval, i.e., proportional to the quantity value.

Degrees of freedom. The concept of degrees of freedom may be explained by choosing a numerical example. Suppose we want to choose two numbers, x and y. There would be an infinite number of choices, n. If, the conditions were that they should add up to 6 there are still an infinite number

TABLE 1 Frequency and cumulative frequency average $= 0$, $s(x) = 1$

Probability	z-Value	Cumul. Area	Frequency
0.001	−3.00	0.001	0.004
0.023	−2.00	0.023	0.054
0.159	−1.00	0.159	0.242
0.309	−0.50	0.309	0.352
0.500	0.00	0.500	0.399
0.691	0.50	0.691	0.352
0.841	1.00	0.841	0.242
0.975	1.96	0.975	0.058
0.977	2.00	0.977	0.054
0.999	3.00	0.999	0.004

for x, or y, to choose from but the choice of one determines the size of the other; the degrees of freedom has been reduced by 1, i.e. $(n-1)$. The value of x depends on y or vice versa, they are not independent.

This can be taken one step further, if $x + y = 6$ and $x^2 + y^2 = 68$ there is only one solution to the equations. Accordingly the $df = 0$. If you have two unknown and two equations the variables can be unambiguously determined.

In statistics this can be generalized to "the number of observations minus the number of necessary relations among these observations" or "the number of measurements with error minus the number of fitted parameters" or "the number of observations minus the number of constraints (c)":

$$df = n - c \tag{16}$$

Degrees of freedom (df) for sample standard deviation: $df = n - 1$ (17)

The subtraction of one from the number of observations (n) in the calculations of the variance and standard deviation of a sample can be intuitively explained by the discussion above; the dataset has been used once already to calculate the average, thus losing one degree of freedom. The algebraic proof of the correction applies to the variance which shows that the bias (underestimation, see below) is corrected in the estimation of the population *variance*, and some, but not all, of the bias in the estimation of the sample *standard deviation* (the square root of the variance). This correction is known as the Bessel's correction.

Standard deviation, shortcut:

This procedure is based on the identity $\sum (x_i - \bar{x})^2 = \sum x_i^2 - \dfrac{\left(\sum x_i \right)^2}{n}$, thus

$$s = \sqrt{\frac{\sum (x_i - \bar{x})^2}{n-1}} = \sqrt{\frac{\sum_{i=1}^{i=n} x_i^2 - \dfrac{\left(\sum_{i=1}^{i=n} x_i \right)^2}{n}}{(n-1)}} = \sqrt{\frac{n \times \sum_{i=1}^{i=n} x_i^2 - \left(\sum_{i=1}^{i=n} x_i \right)^2}{n \times (n-1)}} \tag{18}$$

This formula will give identical results as Eq. (15) but has the procedural advantage of not calculating the average or the individual differences between observations and the average separately. However, both are sensitive to rounding errors and thus sufficient decimals should be included in the calculations and followed by a final correct rounding or the result.

The "sum of squared" values appear in many statistical calculations, e.g., the standard deviation. It is commonly abbreviated SS with a suitable index. The understanding of the term is that the individual values are squared and then added together. There is a special function in EXCEL: *SUMSQ(array)*; in which the referenced array contains the nonsquared values. As indicated in the name, the function yields the sum of the squared items directly. The "sum of squares" will be discussed more in the section "Analysis of variance."

Example: Consider the dataset 5.11, 6.21, 10.98, 5.47, 7.21, 5.90, 3.54, 4.16, 2.69, 4.33, 3.24, $n = 11$. Estimate and compare the standard deviation obtained by the above formulas!

1. (Use the shortcut model). Calculate the differences between each observation and the average (5.35). Square each difference and add them, e.g., using the EXCEL function or apply *SUMSQ(array)* to the differences. The result is 53.53. Divide by the $df = 10$. The variance is 5.35 and the standard deviation is 2.31.
2. Alternatively, calculate the sum of the squared observations = 368.27, the sum of the observations squared = 3462.15, divided by number observations = 3462/11 = 314.74. Calculate the difference (368.27 − 314.74) = 53.53 and divide by the $df = 10$. The variance is 5.35 and the standard deviation = 2.31.
3. There are only 11 observations and it would be erroneous to use the population variance and standard deviation but they can never-the-less be calculated: Population variance: 4.87 and standard deviation 2.21, illustrating the underestimation before Bessel correction.

$$\text{Standard error of the mean: SEM} = s(\bar{x}) = \frac{s(x)}{\sqrt{n}} = \sqrt{\frac{\sum_{i=1}^{i}(x_i - \bar{x})^2}{n \times (n-1)}} \quad (19)$$

The standard error of the mean expresses the interval within which a repeated estimate of the average is assumed to occur with a given probability. The statistic thus expresses how well the average has been estimated.

The SEM will decrease when the number of observations increases in proportion to the square root of the number of observations.

It is important to use the SEM ($s(\bar{x})$) and $s(x)$ correctly to avoid misunderstandings. The standard deviation ($s(x)$) describes the width of the distribution and is thus represented in the definition of the normal distribution; it is theoretically a constant which, together with the average, defines the distribution. It is therefore independent of the number of observations although the more observations made, the better is the estimate of the standard deviation (confidence interval of the standard deviation (Eq. 21)).

The standard error of the mean is abbreviated SEM = $s(\bar{x})$ and with the same logics the standard deviation is abbreviated: $s = s(x)$.

$$\text{Confidence interval of the average: CI} = \bar{X} \pm t_{(1-\alpha;n-1)} \times \frac{s(x)}{\sqrt{n}} \quad (20)$$

The magnitude of the *t*-value depends on the degrees of freedom ($df = n - 1$) and the level of confidence ($1 - \alpha$), CI. The *t*-value should be taken from a Student's *t*-value table or calculated in EXCEL by the function

$T.INV(1 - \alpha, df)$. For a large number of observations (often suggested >30) the t-value can be obtained from a "z-table" or by EXCEL: $NORM.S.INV$ $(1 - \alpha)$, i.e., since the t-distribution approaches the normal distribution and they practically coincide above this number of observations.

Thus, for large number of observations $t_{(1-\alpha, n-1)} \approx z_{(1-\alpha)}$.

To cover a 95 % confidence interval a factor of 2 is often used. This coincides with the "coverage factor" k, applied to the measurement uncertainty.

The confidence interval of the standard deviation is

$$\sqrt{\frac{(n-1) \times s^2}{\chi^2_{(\alpha/2,(n-1))}}} = s \times \sqrt{\frac{(n-1)}{\chi^2_{(\alpha/2,(n-1))}}} \quad \text{to} \quad \sqrt{\frac{(n-1) \times s^2}{\chi^2_{(1-\alpha/2,(n-1))}}} = s \times \sqrt{\frac{(n-1)}{\chi^2_{(1-\alpha/2,(n-1))}}}$$

(21)

where χ^2 is "chi-2" and $(1 - \alpha)$ the confidence level. The degrees of freedom are $(n - 1)$.

Example: In an experiment 25 results were obtained with an average of 5 g and a standard deviation of 0.70 g. Calculate the 95 % confidence interval (CI) for the standard deviation!

First find the χ^2 values for the endpoints of the confidence interval, i.e., 0.975 and 0.025 for the higher and lower limits, respectively. The χ^2 table and EXCEL function $CHIINV(\alpha, df)$ give the part of the distribution to the right (above) of the limit. Thus, for $df = (n - 1)$, i.e., 24, $\chi^2_{(\alpha/2,(n-1))} = 32.9$; $\chi^2_{(1-\alpha/2,(n-1))} = 8.9$.

$$CI_{0.025} = s \times \sqrt{\frac{(n-1)}{\chi^2_{(\alpha/2,(n-1))}}} = 0.7 \times \sqrt{\frac{24}{CHIINV(0.025, 24)}} = 0.547 \quad \text{to}$$

$$CI_{0.975} = s \times \sqrt{\frac{(n-1)}{\chi^2_{(1-\alpha/2,(n-1))}}} = 0.7 \times \sqrt{\frac{24}{CHIINV(0.975, 24)}} = 0.974$$

The standard deviation is thus 0.70 (CI_{95}:0.55−0.97).

NB: The CI of the standard deviation is not symmetrical around the standard deviation.

As illustrated in Fig. 3 the confidence interval of the standard deviation (left panel) is very large if based on few observations. The same tendency is illustrated in a simulated example (right panel) whereas the estimated average (middle panel) levels out within a few observations.

$$\text{Variance sample: } s^2 = \frac{\sum_{i=1}^{i=n} (x_i - \bar{x})^2}{n - 1}$$

(22)

The standard deviation, σ, appears squared in the Gauss formula (36). The standard deviation is easy to visualize but its squared value, the variance, may be more difficult. Variance occurs in many statistical calculations, e.g., propagation of errors and uncertainties and analysis of variance.

The standard deviation, on the average, underestimates the population standard deviations if the number of observations is small (*df*). The sample variance is an "unbiased estimator" of the variance σ^2.

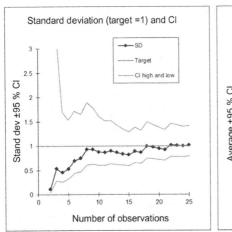
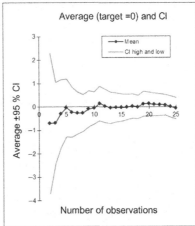

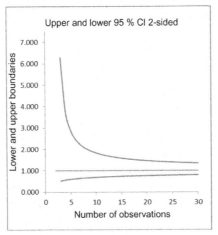

FIGURE 3 A set of 25 consecutive values were drawn from a normally distributed dataset, $N(0;1)$. The average and standard deviation were calculated from the first two, the first three etc. values of the selection. The estimated average is comparatively stable and close to the target after a few observations (middle panel) whereas the standard deviation (left panel) levels out only after about 20 observations. Note the CI is not symmetrical around the estimated standard deviation. The rightmost panel shows the theoretical confidence interval of the standard deviation according to the formula (19).

The expressions "biased" and "unbiased" in this connection refer to the adjusting the naive estimate of the sum of squares with n and $(n-1)$, respectively.

The average underestimation of the population standard deviation by the sample standard deviation σ is about 20 % with 2 observations, 8 % with 4 observations, and about 1 % with 25−26 observations. The average underestimation can be corrected by the *c4 correction* based on the gamma function, e.g., in EXCEL:

$$c4 = EXP\left(LN\left(SQRT\left(2/(N-1)\right)\right) + GAMMALN\left(N/2\right) - GAMMALN\left((N-1)/2\right)\right)$$

(23)

Number of observations	2	3	4	5	10	25	26	100
c4		0.7979	0.8862	0.9213	0.9400	0.9727	0.9896	0.9901	0.9975

The correction is achieved by dividing the estimated $s(x)$ by the appropriate table value.

Example: A series of results was 3, 3, 6, 2, 7, 5, 9, 3, 8, 4. The average is 5 and the standard deviation 2.28. The % CV is then $100 \times 2.28/5 = 45.6$ %.

NB: The relative standard deviation (RSD) and the coefficient of variation are always (by definitions (28) and (29)) calculated from one standard deviation.

Standard deviation, pooled:

$$s_{\text{pool}} = \sqrt{\frac{(n_1 - 1) \times s_1^2 + (n_2 - 1) \times s_2^2 + \cdots + (n_k - 1) \times s_k^2}{(n_1 + n_2 + \cdots + n_k) - k}}$$

(24)

where indices *1,2 ... k refers to the number of observations (n) and standard deviations(s) in different experiments.*

The pooled standard deviation is used to find an estimate of the sample standard deviation given several different samples and their number and standard deviation measured under different conditions. The average of the different series of measurements may vary between samples but the standard deviation (imprecision) should remain almost the same. Pooling standard deviations may provide an improved estimate of the imprecision and is the weighted average of many estimates of standard deviations.

If the number of observations is the same in each sample $(n_1 = n_2 = \cdots = n_i)$, then

$$s_{pool} = \sqrt{\frac{(n_1 - 1) \times s_1^2 + (n_1 - 1) \times s_2^2 + \cdots + (n_1 - 1) \times s_k^2}{k \times n - k}}$$

$$= \sqrt{\frac{s_1^2 + s_2^2 + \cdots + s_k^2}{k}} = \sqrt{\frac{\sum_{i=1}^{k} s_i^2}{k}} \tag{25}$$

Since the number of observations is the same, the weighting factor is one (1) and the pooled $s(x)$ can be thought of as the average of standard deviations in an experiment i.e., the s_{pool} is the square root of the average *variance* of the groups.

The pooled standard deviation can be used to describe the total performance of a laboratory which uses many instruments to measure the concentration of the same analyte.

Example: A laboratory used three instruments to measure the concentration of the same analyte. The first instrument measured 12 samples with a standard deviation (s) of 0.35, the second 15 samples with $s = 0.55$ and the third instrument measured seven samples with $s = 0.40$. Estimate the pooled standard deviation, representing that for samples analyzed on a random choice of instruments.

$$s_{pool} = \sqrt{\frac{(12 - 1) \times 0.35^2 + (15 - 1) \times 0.55^2 + (7 - 1) \times 0.40^2}{(12 + 15 + 7) - 3}}$$

$$= \sqrt{\frac{6.54}{34 - 3}} = \sqrt{0.21} = 0.46$$

Squared, multiplied by $(n - k)$ and applied to repeated measurements of the same material, this is identical to the "within group" or "Average Square" of the ANOVA; compare the formula for within series Sum of Squares in ANOVA analysis, SS_w (Eq. 125).

If the imprecision is not constant but proportional to the measured quantity it is reasonable to assume that there is a constant relative standard deviation, characterizing the imprecision. The calculated relative standard deviation can be calculated from the results of a series of measurements as

$$RSD_{pool} = \sqrt{\frac{\left(\sum_{i=1}^{k} \frac{(n_i - 1) \times s_i^2}{\bar{x}_i^2} \right)}{\sum_{i=1}^{k} n_i - k}} \tag{26}$$

Standard deviation (Dahlberg):

If there are only two observations in each group then Eq. (30) is rearranged to:

$$s = \sqrt{\frac{\sum (n-1) \times (x_i - \bar{x})^2}{\sum n_i - k}} = \sqrt{\frac{\sum (2-1) \times \left(x_i - \frac{x_{i1} - x_{i2}}{2} \right)^2}{k \times (2-1)}} = \sqrt{\frac{\sum (x_{i1} - x_{i2})^2}{4 \times k}}$$

Finally, since $2 \times k = N$,

$$s = \sqrt{\frac{\sum (x_{i1} - x_{i2})^2}{4 \times k}} = \sqrt{\frac{\sum_{i=1}^{N} (x_{1i} - x_{2i})^2}{2 \times N}} = \sqrt{\frac{\sum_{i=1}^{i-N} d_i^2}{2 \times N}} \qquad (27)$$

where d is the difference between results of duplicate measurements, N is the number of pairs.

This method therefore estimates the standard deviation from a number of duplicate measurements and is used to estimate the imprecision of a particular measurement procedure when duplicate measurements are available. It is important that the standard deviation is constant in the measuring interval (homoscedastic).

If instead the relative standard deviation (RSD) is constant, i.e., the standard deviation is proportional to the quantity value (heteroscedastic), the RSD from duplicate observations is calculated from the relative difference between each pair of:

$$\text{RSD} = \sqrt{\frac{\sum_{i=1}^{i=N} \left(\frac{2 \times (x_{i1} - x_{i2})}{(x_{i1} + x_{i2})} \right)^2}{2 \times N}} \qquad (28)$$

The difference is thus calculated in relation to the average of the two observations. The relative standard deviation cannot be calculated in relation to the average of all observations since the distribution of the duplicate values is not necessarily normal.

Example: A series of samples were measured in duplicates: 3.6−4.0; 10.2−9.8; 5.1−5.9; 4.6−4.2; and 7.2−7.6. Calculate the standard deviation of the procedure!

The differences between the duplicates were: −0.4; 0.4; −0.8; 0.4; and −0.4. The sum of the squared differences $\sum d^2 = 1.28$; $s = \sqrt{\frac{1.28}{2 \times 5}} = 0.36$.

Similarly the RSD:

$$\text{RSD} = \sqrt{\frac{\left(\frac{-0.4}{3.8} \right)^2 + \left(\frac{0.4}{10.0} \right)^2 + \left(\frac{-0.8}{5.5} \right)^2 + \left(\frac{0.4}{4.4} \right)^2 + \left(\frac{-0.4}{7.4} \right)^2}{2 \times 5}}$$

$$= \sqrt{\frac{0.045}{2 \times 5}} = 0.0671 \text{ or } 6.7\%.$$

The % CV calculated from the s and related to the average of 6.22, would be $100 \times \dfrac{0.36}{6.22} = 5.6\,\%$.

In the example we have no information on the homo- or heteroscedasticity of the measurements and it illustrates the importance of avoiding the overall average for the % CV.

Standard deviation, geometric is obtained by calculation of the standard deviation from logarithmically transformed measurement results $\bar{x}_g$ and s_g, respectively. The logarithmic confidence interval is

$$\mathrm{CI}_g = \bar{x}_g \pm z \times \frac{s_g}{\sqrt{n}} \tag{29}$$

To get the quantity values the antilogarithms of $\bar{x}_g$ and s_g must be calculated.

Root mean square (mean, quadratic; RMS)

$$x_q = \sqrt{\frac{\sum_{i=1}^{i=n} x_i^2}{n}} = \sqrt{\frac{1}{n}\left[(x_1)^2 + (x_2)^2 + \dots(x_n)^2\right]} \tag{30}$$

The RMS is a measure of the magnitude of a varying quantity. It is especially used when variables vary between positive and negative values of the same magnitude, e.g., in describing a sinus wave. It corresponds to the standard deviation of a number of observations with an average of zero (cf. Eq. 14); if the standard deviation describes the variation of normally distributed data around an average, the RMS describes the variation around zero.

$$\text{Mean squared error: MSE} = \frac{\sum_{i=1}^{n}(x_i - x_0)^2}{n} \tag{31}$$

where x_0 is a true or a predetermined value according to a model. The MSE is frequently used in business analysis where the x_0 is the predicted value and the MSE taken as a measure of reliability.

The RMS of the Deviation of a Measurement

The "root mean square of the deviation of a measurement" (Δ) is also a statistic which is used in forecasting and the basis of quality management in clinical laboratories, particularly in Germany, under the name of Rili-BAEK. It is a metric that addresses the imprecision *and* bias. It is the average of the squared deviations from a target value. If the target value is equal to the average of the distribution then it equals the population standard deviation (Eq. 14) of the distribution, if the target value is zero it equals the root mean square (Eq. 24) and equals cyclic variation.

$$\Delta = \sqrt{\frac{\sum_{i=1}^{i=n}(x_i - x_0)^2}{n}} \tag{32}$$

where x_i is the observations and x_0 a "target value."

If the average of the observations and their standard deviation are $(\bar{x})$ and (s), respectively, and the systematic deviation $(\bar{x} - x_0) = \delta$ then Eq. (26) can be rewritten, since $\sum_{i=1}^{n}(x_i - x_0)^2 = \sum_{i=1}^{n}(x_i - \bar{x})^2 + n \times (\bar{x} - x_0)^2$, as

$$\Delta = \sqrt{\frac{\sum_{i=1}^{i=n}(x_i - x_0)^2}{n}} = \sqrt{\frac{\sum_{i=1}^{i=n}(x_i - \bar{x})^2 + n \times (\bar{x} - x_0)^2}{n}} = \sqrt{s_p^2 + (\bar{x} - x_0)^2} =$$

$$= \sqrt{\frac{n-1}{n} \times s_s^2 + \delta^2} \tag{33}$$

It is noteworthy that the Δ includes both bias and random variation (obvious from Eq. (27)) and therefore allows an infinite number of combinations of the two.

$$\text{Standard deviation, relative (coefficient of variation): } CV = \frac{s(x)}{\bar{x}} \tag{34}$$

$$\text{Standard deviation, relative, percent: } \%CV = \frac{100 \times s(x)}{\bar{x}} \tag{35}$$

The coefficient of variation, percent, is often abbreviated % CV but other abbreviations are also common.

2.3 The Normal and *t*-Distributions

2.3.1 The Normal Distribution

Distributions of data are referred to as "probability density functions." The Gaussian or normal, distribution, is a symmetrical distribution of the data around the average, with a width defined by the standard deviation. Therefore, the average places the distribution along the number line and the standard deviation expresses the dispersion of the data. The Gaussian distribution represents the expected distribution of random data, for instance from nonbiased repeated observations.

$$\text{Gaussian distribution: } N(\mu, \sigma) = \frac{1}{\sqrt{2\pi\sigma^2}} \times e^{\frac{-(x_i - \mu)^2}{2\sigma^2}} = \frac{1}{\sigma\sqrt{2\pi}} \times e^{\left(\left(-\frac{1}{2}\right) \times \left(\frac{x_i - \mu}{\sigma}\right)^2\right)} \tag{36}$$

The conventional notation for a Gaussian distribution is $N(\mu, \sigma)$.

The distribution is fully defined by two characteristics of the data, the population average, μ, and the population standard deviation, σ. The only variable in the expression is the x_i which is expressed on the X-axis and has a corresponding probability, shown at the intersection between a vertical line through x_i, and the frequency distribution $N(\mu,\sigma)$ in EXCEL: NORM.S.DIST (z,false). Thus, for each x_i the expression gives the corresponding frequency, i.e. Y-value.

The normal distribution is graphically represented by the well-known bell-shaped curve, residing on a horizontal value axis (X-axis), and the frequency of observations on the vertical Y-axis. The peak of the curve is the average of all observations of the population and the width of the distribution is the standard deviation. See Fig. 2.

The frequency of the Gaussian function (y_i) (36) can be coded in EXCEL at any X-value (x_i) by the following:

$$y_i = \frac{1}{s\sqrt{2\pi}} \times EXP\left(-\frac{1}{2} \times \left((x_i - \bar{x})/s\right)^2\right)$$

A "standard normal distribution" $N(0,1)$ is characterized by an average of 0 and a standard deviation (variance) of 1. Formula (36) is then simplified to:

$$N(0,1) = \frac{1}{\sqrt{2\pi}} \times e^{-\left(\frac{(x_i)^2}{2}\right)} \tag{37}$$

As a result of the symmetry of the distribution the area under the curve (AUC) is the same on both sides of the axis of symmetry, i.e. through the average, in this case the Y-axis. A quantity value and thus the frequency value, below the average has a counterpart above.

Any quantity value in the distribution can be expressed in relation to the standard deviation, "the standardized deviate." This relative value is the z-value,

$$z_i = \frac{x_i - \mu}{\sigma} \tag{38}$$

The probability of a value is the cumulated AUC from $-\infty$ to $\mu + z$, known as the cumulative probability. The z-value can be any number, negative below the average and positive above.

For instance, if $z_1 = -1.96$ and $z_2 = 1.96$ the AUC in the interval $\mu + z_1$ to $\mu + z_2$ includes about 97.5 % of the distribution, see Table 1.

For a standard normal distribution the probability is coded in EXCEL as NORM.S.DIST(z,true). In previous versions of EXCEL, before 2007, the function was NORMSDIST(z).

To describe an arbitrary normal distribution ($N(\bar{x};s)$) the average and standard deviation are required, thus NORM.DIST(z,average,standdev,true). The "true" is to generate the cumulative frequency.

The functions $NORM.DIST(x_i, average, standdev, false)$ and $NORM.S.DIST$ $(z, false)$ give the frequency value of the normal distribution function, i.e., the value of y, at the given value of z or x_i, respectively. The options *false* and *true* give the possibility to directly construct the density and the cumulative probability functions, see Table 1.

Calculation of the standard deviation (expressed as z-value) at which a certain probability (AUC) is reached, is coded as $NORM.S.INV(probability)$ for a standard normal distribution and $NORM.INV(probability, average, standarddev)$ for the general probability function. Thus, $NORM.S.INV(0.5) = 0$, i.e., the value of the average, where the probability is 50 % (0.5). Consequently, the average $+1 \times s(x)$ represents a probability increase of $(0.841 - 0.500) = 0.341$, "the second $s(x)$" $(0.977 - 0.841) = 0.136$ and the average $2 \times s(x) = 2 \times (0.977 - 0.500) = 0.950$. $3 \times s(x)$ is usually approximated to a probability of 99.9 % and $4 \times s(x)$ to 99.99 %. See Table 2.

2.3.2   To Draw Normal Distribution Curves in EXCEL

By defining the average and standard deviation and specifying a suitable interval of quantity values, the formula (36) can be used to calculate the frequency at each of defined quantity values. The generated table can be displayed graphically and thus show the typical bell-shaped curve.

Although the basic shape (a symmetric "bell" with infinite "tails") is the same, the distribution's width and height may vary, also in relation to each other. There are two concepts to describe this, skewness and kurtosis.

2.3.3   Skewness

Skewness is a measure of the asymmetry of the probability distribution of a random variable. The skewness can be expressed numerically and is usually

TABLE 2 The relation between the probability (Cumulative AUC) and the z-Value

Cumulative AUC (Probability)	z-Value	Above the z-Value (1-AUC)	Cumulative AUC (Probability)	z-Value	Above the z-Value (1-AUC)
0.025	−1.96	0.975	0.977	2.00	0.023
0.159	−1.00	0.841	0.99000	2.33	0.01000
0.500	0.00	0.500	0.99900	3.09	0.00100
0.841	1.00	0.159	0.99990	3.72	0.00010
0.975	1.96	0.025	0.99999	4.27	0.00001

displayed in statistics' software. A positive skewness indicates that the "tail" of the distribution is extended towards higher values (a "right skewness"), still the majority of observations may be found at values below the calculated average. Likewise, a negative skewness indicates a "left" skewness with a tail extending to the left and the majority of observations concentrated to high values.

A normal distribution of a continuous variable has a skewness of 0. The "standard" deviation of a skew distribution, may still be calculated from the formulas, but can no longer be interpreted as described above.

There are many ways to numerically estimate and express the skewness. Statistically it was defined in terms of the second and third "moment" about the average (Pearson) as

$$g = \frac{\frac{1}{n} \times \sum_{i=1}^{n} (x_i - \bar{x})^3}{\left[\frac{1}{n} \times \sum_{i=1}^{n} (x_i - \bar{x})^2\right]^{3/2}}$$

This formula can be expanded to compensate for the sample size:

$$G = \frac{\sqrt{n \times (n-1)}}{(n-2)} \times \frac{\frac{1}{n} \times \sum_{i=1}^{n} (x_i - \bar{x})^3}{\left[\frac{1}{n} \times \sum_{i=1}^{n} (x_i - \bar{x})^2\right]^{3/2}} = \frac{n}{(n-1) \times (n-2)} \times \sum_{i=1}^{n} \left(\frac{(x_i - \bar{x})^2}{s}\right)^3$$

(39)

which is the formula used in the EXCEL *SKEW(array)* function.

There are tables of critical values of G for normal distributions.

By convention the skewness calculated according to Eq. (38) is interpreted as:

- less than -1 or greater than $+1$: highly skewed;
- between -1 and $-\frac{1}{2}$ or between $+\frac{1}{2}$ and $+1$: moderately skewed;
- between $-\frac{1}{2}$ and $+\frac{1}{2}$: approximately symmetric.

Other measures of the skewness are based on the observation that the average, median, and mode coincide in a symmetrical distribution, whereas in a nonsymmetrical distribution they do not. Symmetrical distributions can therefore pass one of these tests but still not be Gaussian.

The Bowley skewness or quartile skewness considers the median and quartiles.

$$S = \frac{q_3 - q_2 - (q_2 - q_1)}{q_3 - q_1} = \frac{q_3 + q_1 - 2 \times q_2}{q_3 - q_1} = p(0.25) + p(0.75) - \frac{2 \times p(0.50)}{p(0.75) - p(0.25)}$$

(40)

where q_3 is the third quartile, $p(0.75)$, and q_1 the first quartile, $p(0.25)$, and q_2 is the median, $p(0.50)$. In this formula S takes the values -1 to $+1$. In a symmetrical distribution $S = 0$, if positive there is a right skewness and if negative a left skewness.

Since only the middle and two quartiles of the distribution are considered, and the outer two quartiles are ignored, this adds robustness to the measure but is also a caveat. The choice of quartiles as the limits is arbitrary and other limits may be preferred, e.g., $p(0.05)$ and $p(0.95)$. Therefore the quartile skewness needs to be further specified, in practical use.

The Pearson skewness is

$$S_1 = \frac{\bar{x} - \text{mode}}{s(x)} \tag{41}$$

and the Pearson second skewness

$$S_2 = \frac{\bar{x} - \text{median}}{s((x))} \tag{42}$$

The "second skewness" may be more practical since the median can easily be calculated and the distribution may be multimodal. The Pearson first and second skewness approaches are to express the difference between the average and the median, or mode, in terms of the standard deviation estimated as if the distribution were normal.

The "Pearson skewness index" is

$$S_{kp} = \frac{3 \times (\text{average} - \text{median})}{s(x)} \tag{43}$$

Example: Suppose we have a dataset that is normally distributed with an average of 0 and a standard deviation of 1. Characteristically the median, average, and mode are equal. Suppose further that we add a number of results that are much larger than the average and an equal number of results smaller than the average but close to the average. This causes the average to move to a higher value and the standard deviation to increase, whereas the median remains the same. Consequently, the Pearson skewness turns positive.

An extensive example of skewness is given below.

2.3.4 Kurtosis

Kurtosis is a measure of the "tailedness" of the probability distribution. A standard normal distribution has kurtosis of 3 and is recognized as mesokurtic. An increased kurtosis (>3) can be visualized as a thin "bell" with a high peak whereas a decreased kurtosis corresponds to a broadening of the peak and "thickening" of the tails. Kurtosis >3 is recognized as leptokurtic and <3 as platykurtic (lepto = thin; platy = broad). There are four different formats of kurtosis, the simplest is the population kurtosis; the ratio between the fourth moment and the variance.

In EXCEL the "excess kurtosis" is calculated by the function *KURT (array)* which gives the population kurtosis minus 3 (kurtois-3). Therefore, in EXCEL zero indicates a perfect tailedness and positive values a leptokurtic distribution.

A detailed knowledge of skewness and kurtosis is rarely important in the laboratory but since the statistics are easily available they may serve as indicators of normality.

2.3.5 The t-Distribution

The population average, μ, and the population standard deviation, σ, define the normal distribution $N(\mu,\sigma)$. The probability for any particular value is calculated from the z-value (Eq. 37). If the standard deviation (σ) is not known but the standard deviation of the sample mean, estimated as $s/\sqrt{n}$, i.e., the SEM, the "normalized deviate" (t) is

$$t = \frac{\bar{x} - \mu}{s/\sqrt{n}} \tag{44}$$

When the number of observations is low, i.e., the population parameters are estimated from a sample, the number of observations must be taken into account. It is also the situation when the σ is not known and a Bessel correction for the biased standard deviation, i.e., n − 1 should be used. Using the t as the normalized deviate leads to the *t-distribution*, attributed to Gosset, known as "Student" in the literature.

The term "normalized deviate" indicates that a t-value can be calculated and interpreted for any difference which is normally distributed. Usually we recognize and differ between Student's independent (Eq. 94) and dependent (Eq. 99) t-values. Understanding the "normalized deviate" concept clarifies that the distribution of the observations and the distribution of the differences between individual observations, respectively, need to be normally distributed in the two calculations.

The obtained *t-distribution* basically has the same bell-shaped symmetrical form as the normal distribution but is wider, less peaked; the exact shape depending on the degrees of freedom (Eq. 16). The *t-distribution* approaches that of the normal when the number of observations increases. To qualify this statement the table values of the t-distribution for different degrees of freedom are shown in Table 3.

2.4 Transformation of Distributions

Many statistical evaluations require that the distribution of the data is, or is close to, normal. A dataset can sometimes be transformed to approach a normal distribution by recalculating the quantity values to logarithms, the square

TABLE 3 The total probability of values of the normal distribution outside 3s is 0.0027, ((1−NORM.S.DIST(3,true)) × 2) and independent of the *df* *whereas the probability outside 3s in a t-distribution is shown on the* *second line (((1-T.DIST(3;true)) x 2)*

df	2	5	10	20	30	100
p	0.0955	0.0301	0.0133	0.0071	0.0054	0.0034

TABLE 4 Descriptive statistics of the original and transformed distributions

Transformed	Orig	Ln	Sq Root
Average	2.57	0.87	1.60
Median	2.20	0.79	1.48
Stdev	1.08	0.38	0.32
Confidence interval	± 0.12	± 0.04	± 0.03
Mode	1.70	0.53	1.30
Percentile 75	3.10	1.13	1.76
Percentile 25	1.77	0.57	1.33
Recalculated	**Orig**	**Anti-Ln**	**Squared**
Average	2.57	2.39	2.56
Median	2.20	2.20	2.19
Stdev	1.08	1.46	0.10
Average − stdev	1.49	0.93	1.58
Average + stdev	3.66	3.85	3.57
CI (± SEM, n = 84)	2.45−2.69	2.29−2.48	2.46−2.66
Mode	1.70	1.70	1.70
Percentile 75	3.10	3.10	3.10
Percentile 25	1.77	1.77	1.77

roots, or reciprocal values. These techniques decrease large values more than small values and positively skewed datasets may come close to normal. In general terms reciprocal ($1/x_i$) has a stronger effect on the skewness than the logarithmic transformation and the square root a weaker effect (see Table 4).

The Box-Cox transformation is essentially an exponential transformation and a system to estimate an optimal exponent between -5 and $+5$.

Transformation of data may facilitate the calculation of descriptive statistics but it is usually not good practice to report summary results before reestablishing or "back-transforming" the results. This is done by applying the opposite mathematical operation to that used in the transformation.

Example: The concentration of triglycerides in a series of 84 patient samples was measured.

Two transformations of the raw data were tried, logarithmic and square root (Fig. 4). The descriptive statistics of the transformed distributions were transformed back to the original format. The averages have shifted and the confidence interval is no longer symmetrical around the estimated average; the nonparametric quantities (e.g., median, percentiles) are unchanged. Both

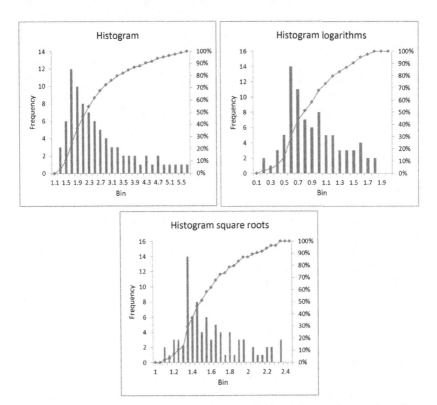

FIGURE 4 The appearance of a skew frequency distribution (left) and after transformations. The display of the distribution is related to the choice of bin sizes, whereas the skewness indicators are independent.

TABLE 5 Comparison of Skewness estimates

	Orig	Ln	Sq Root
Skew (EXCEL)	1.16	0.49	0.83
Skew (Bowley)	0.36	0.23	0.29
Skew (Pearson-2)	0.34	0.21	0.28

transformations resulted in a reduction of the skewness (Table 5). As illustrated in Table 5, logarithmic transformation is more powerful than the square root transformation.

NB: Many procedures which are used in the laboratory assume a Gaussian distribution of the data. There are various methods to test normality (e.g., see the section "Quantile") but there are some rules of thumb which might be useful to consider. This is particularly important since you will always get a result if available formulas are used but the results cannot be interpreted. Thus, in a Gaussian distribution the average and the median coincide, kurtosis is close to 3 (i.e., 0 in EXCEL), skewness is below the absolute value 1. For a physical quantity the distribution is always positive (e.g., a concentration cannot be negative). Therefore, the average minus one, two, or three times the standard deviation must still be positive. Further discussions of normality will be under the section "Robust estimators."

2.5 Random Numbers and Simulation of a Normal Distribution

Random numbers can be obtained from tables or number generators (e.g., www.random.org) but can also be generated using EXCEL through a "Pseudo random number generator" (PRNG). The function *RAND()* will return one single random number between 0 and 1. If repeated the cumulative data will represent a rectangular distribution. A new random number will be generated every time the F9 button is pressed.

The random choice of a number between 0 and 1 which is returned by the function *RAND()* can be used to create a set of normally distributed data in EXCEL. The "probability" in the function *NORM.S.INV(probability)* should then be substituted by *RAND(): NORM.S.INV(RAND())*. If this formula is repeated in a column or row as an array, a set of normally distributed random numbers with an average of 0 and standard deviation of 1 is obtained. Adding the average ($\bar{x}$) and the standard deviation $s(x)$ will modify the location and width of the generated distribution: $\bar{x} + s(x) \times NORM.S.INV$ *(RAND())*.

Another function *NORM.INV(RAND(),average,s(x))* can also be used to generate a set of normally distributed values with the given average and standard deviation.

Two algorithms use the relation of the PRNG to a rectangular distribution in the generation of normal distributions:

$$\bar{x} + s(x) \times (RAND() + RAND() + RAND() + RAND() + RAND() + RAND() + RAND() + RAND() + RAND() + RAND() + RAND() + RAND() - 6)$$

and

$$\bar{x} + s(x) \times (RAND() - RAND() + RAND() - RAND() + RAND() - RAND() + RAND() - RAND() + RAND() - RAND() + RAND() - RAND()).$$

There is also a function "Random number generator" in the "Data analysis" add-in which allows simulation of different types of distributions, including the Gauss distribution.

The use of the PRNG in different algorithms renders slightly different properties to the created normal distributions. The "quality" of the obtained function is depending on the PRNG and the other functions. For instance small differences are obtained between distributions generated by EXCEL and *R*.

3. PERMUTATIONS, COMBINATIONS, PROBABILITY, BINOMIAL, AND POISSON DISTRIBUTIONS

In evaluating the possible number of combinations in a population or sample, different conditions may apply and strict descriptions of the conditions are needed. The mathematical/statistical definitions of the conditions may sound awkward to the layman but this is the terminology:

Permutation is *arranging* a set of items in a certain order. Thus, order matters.

Combination is *selecting* a set of items independent of the order in which the items occur; order does not matter.

Possible combinations will be different depending on if repetition of samples is allowed or not.

In the following examples n will be the total number of available items and r the number selected.

Permutations with repetition, e.g., the code of a "combination lock" in which the same number (digit) can be used many times, i.e., independent of previous use. The total number of combinations will be

$$P(n, r) = n^r \tag{45}$$

Example: There are 10 digits 0, 1, ..., 9 and the code consists of for instance three digits. You can argue that in the first position the numbers can be selected in 10 ways, in the second and third positions also 10 ways. The total is the 10 *and* 10 *and* 10 ways. "AND" meaning multiplication ($\cap$) and the total number of possible permutations, thus $n^r = 10^3 = 1000$.

In this scenario all possible permutations will be limited by the ordering and the results may only be used once.

Permutation without repetition:

Example: Think of estimating how many gold, silver, and bronze medal combinations there could be in a group of competitors. Once you have found one it must not be considered again. Thus the number will be reduced in relation to when repetitions were allowed. In the first position (gold) you can chose from 10 competitors, for the second you have only nine candidates left and for bronze only eight.

$$P(n,r) = {}^nP_r = {}_nP_r = \frac{n!}{(n-r)!} = \frac{10!}{(10-3)!} = 720 \text{ possible combinations}$$
of medal winners!

Using EXCEL: *FACT(10)/FACT(7)*

Combination with repetition:

Example: There are five types of sweets to choose from and three may be chosen to each bag. How many different bags of sweets are there?

The bags may contain all the same or different sweets, repetition is allowed and the order is irrelevant.

$$C(n,r) = \binom{n+r-1}{r} = \binom{n+r-1}{n-1} = \frac{(n+r-1)!}{r! \times (n-1)!} = \frac{(5+3-1)!}{3! \times (5-1)!} = 35$$

This is the usual situation in a lottery. A lottery is a game in which you shall pick a certain number of digits, i.e., ticket, out of a set of tickets (numbers).

Combination without repetition:

Example: How many unique sets of three digits, in any order, out of 50 available will there be?

$$C(n,r) = {}^nC_r = {}_nC_r = \binom{n}{r} = \binom{n}{n-1} = \frac{n!}{r! \times (n-r)!} = \frac{50!}{3! \times (50-3)!} = 19,600$$

$$(46)$$

3.1 Inclusive and Exclusive Events

There are other situations which need to be considered. The permutation with repetition can be expanded to cover different probabilities for each event, e.g., $p(x)$ and $p(y)$. Provided the events are *independent*, the $p(x$ and $y)$ will be $p(x) \times p(y)$.

Example: What is the probability of getting five 6s *(or any other predetermined number)* in a row (or in five dice thrown at the same time). The probability is the same (1/6) in each dice, i.e., $(1/6)^5 = 1/7776$.

If the events are *dependent*, the combined probability is the product of the pre- and post-probabilities.

Example: What is the probability to draw two black cards from a deck of cards?

The probability to draw a black card first hand is 26/52, if that happened the probability to draw another is 25/51, otherwise it is 26/51. The combined probability is therefore $p(1st) \times p(2nd)$.

If two events are mutually exclusive, i.e., cannot happen together, one *or* the other event will take place. Therefore the probabilities are added.

Example: Suppose there are eight differently colored buttons, two are red and one is blue, the rest are green or yellow. If a person presses (randomly) the red or blue buttons they will win. Probabilities of pressing the red button $p(red)$ will be 2/8 and $p(blue)$ will be 1/8. The probability of winning is $p(red) + p(blue)$ is 3/8.

Events which are mutually inclusive can happen at the same time.

Example: What is the probability of getting a result >3 or a multiple of 2 (2, 4, or 6) when throwing a dice?

More than $(>)3$: 3/6 or [2,4,6]: 3/6 or [4 and 6]: $-2/6$; therefore $3/6 + 3/6 - 2/6 = 4/6 = 2/3$.

3.2 Probability and Odds

The ratio between the number of studied events with a "positive" result that occurred (r) and the total number of events (n) is the probability (p) of the event happening. If the probability is p, the probability of the opposite event (negative) is $1 - p = q$. Small letters are used to symbolize "sample"

$$p = \frac{r}{n} \qquad (47)$$

Colloquially, we are familiar with *risk* (e.g., the risk for cancer), *chance* (to win in a lottery), and *odds* (for a horse to win a race) to describe the probability.

Particularly *odds* and its relation to *probability* may need closer attention. If odds *for* is abbreviated O_f and *against* O_a then,

$$O_f = \frac{p}{q} = \frac{p}{1-p} = \frac{1-q}{q} \quad \text{and } O_a = \frac{q}{p} = \frac{1-p}{p} = \frac{q}{1-q} \tag{48}$$

In a classical explanation, consider a series XXXXYY, i.e., six events where X is "failure" and Y is "success," independent of the value of the words and could also be 0 and 1, respectively.

The probability of success, i.e., Y taking place is YY/XXXOXYY = 2/6 = 1/3 whereas the odds would be YY/XXXX = 2/4 = 1/2 (for) or XXXX/YY = 4/2 = 2/1 (against). Unless "for" or "against" is specified, the odds can be misunderstood.

Since $O_f = \dfrac{p}{1-p}$ a high probability *for* corresponds to high odds. This is opposite to what high odds usually mean at the race course where instead high odds indicate a low probability that the particular horse will win, i.e., $O_a = \dfrac{1-p}{p}$.

Probability can take any number between 0 and 1 whereas odds can be anything between 0 and infinity (∞).

In summary:

Probability is the number of times success occurred compared to the total number of trials.
Odds are the number of times success occurred compared to the number of times failure occurred.

3.3   Binomial Distribution

When an event can have only two outcomes, e.g., *yes/no*, this is a binary event. Tossing a fair coin, the "Bernoulli trial," will result in heads and tails at the same frequency, i.e., p for either heads or tails is 0.5.

The "binominal experiment" refers to the probability of obtaining a particular outcome in a series of repeated trials. The probability ($p(r)$) of obtaining exactly r successes in n trials in a sample with a probability of p is

$$p(r) = \frac{n!}{r! \times (n-r)!} \times p^r \times (1-p)^{(n-r)} \tag{49}$$

The left part of the expression, $\dfrac{n!}{r! \times (n-r)!}$ is known as the binominal

coefficient and written as $\begin{pmatrix} n \\ r \end{pmatrix}$ (pronounced "n choose r") or $C(n,r)$. This corresponds to the number of possible combinations without repetition (Eq. 45). The factor p^r represents the probability of success in a series (r) of trails trial and $(1-p)^{(n-r)}$ the probability of failure in the same series of trials.

The value of the binominal coefficient can be obtained from Pascal's triangle for each value of n and r.

Example: If the Bernoulli trial above were repeated 10 times what is the probability of exactly four "heads"?

$$p(r) = \begin{pmatrix} 10 \\ 4 \end{pmatrix} \times 0.5^4 \times (1-0.5)^{(10-4)} = \frac{10!}{4! \times (6)!} \times 0.5^4 \times (0.5)^6 = 0.21$$

i.e., the *odds* $O_f = \dfrac{0.21}{1-0.21}$ expressed as 21:79, i.e., 1:3.8.

NB: To interpret "odds" in terms of probabilities these must be transferred back to probability using the appropriate expression in Eq. (47).

The "binominal experiment" can be solved using a function in EXCEL: *BINOM.DIST(number of successes,number of trials,probability,cumulative [true or false])*.

Information about a "probability for a success" can be used to extrapolate to another situation.

Example: The error rate in a measurement procedure is 5 %. Estimate the expected number of errors in a series of 200 measurements with a 95 % confidence!

Use the EXCEL BINOM.INV*(trials,probability,confidence)*. The expected number is 15.

Standard deviation of a proportion $(p/(1-p)$: $s\,(p) = \sqrt{p \times (1-p)}$ (50)

The frequency distribution of a proportion is the binominal distribution. The normal distribution is a good approximation since the binominal distribution approaches the normal distribution when the number of events increases. It is usually deemed sufficient if both $n \times p$ and $n \times (1-p)$ exceed 5 which is equal to requiring that r and $n-r$ are above 5.

Standard error of a ratio (probability [p]) $s\,(\overline{p}) = \sqrt{\dfrac{p \times (1-p)}{n}}$ (51)

where p is the proportion and n is the number of observations i.e., the sample size.

The confidence interval $\mathrm{CI}(p) = z \times \sqrt{\dfrac{p \times (1-p)}{n}}$ (52)

The above method is often referred to as the traditional method to estimate $s(\bar{p})$. It is not applicable if the proportion is large or small; often limits of 0.1 and 0.9 are specified. It is thus often inappropriate to use the traditional method for estimating the standard error and confidence interval of diagnostic sensitivity, diagnostic specificity, and prevalence of disease, where probabilities below and above these limits, respectively, are common. Instead, the following procedure is recommended, often referred to as Wilson's method.

If r is the number of observations that refers to a particular property (e.g., testing positive true positive TP, in an investigation among diseased), in a sample of n observations (e.g., diseased), the probability of finding r is obtained in a simplified approach:

$$p = \frac{r}{n} \text{ and } q = 1 - p$$

Calculate:

$$A = 2 \times r + z^2 \tag{53}$$

$$B = z \times \sqrt{z^2 + 4 \times r \times q} \tag{54}$$

$$C = 2 \times (n + z^2) \tag{55}$$

where z is 1.96 to correspond to a 95 % confidence interval.
Then, the confidence interval is

$$\frac{A - B}{C} \text{ to } \frac{A + B}{C} \tag{56}$$

The confidence interval according to Wilson and described above can be written in one expression as

$$\pm \mathrm{CI} = \frac{p + \dfrac{z^2}{2 \times n} \pm \sqrt{\dfrac{p \times (1-p)}{n} + \dfrac{z^2}{4 \times n^2}}}{1 + \dfrac{z^2}{n}} \tag{57}$$

where the $\pm$ sign indicates the upper and lower limits.

None of the methods to estimate the confidence limits of a proportion is applicable in all situations and there are many procedures available. The Wilson score is supposed to give a reasonably conservative confidence interval.

Example: In a group of 85 (n) known diseased only five tested positive (r) with a new test. Estimate the proportion (p) of true positives among all diseased (TP = *sensitivity*) and its confidence interval using traditional and alternative methods!

Thus, $n = 85$ and $r = 5$ and $p = \frac{5}{85} = 0.059$; $n \times (1 - p) = 80$; $n - r = 75$

$$A = 2 \times 5 + 1.96^2 = 13.84$$

$$B = 1.96 \times \sqrt{1.96^2 + 4 \times 5 \times \left(1 - \frac{5}{85}\right)} = 9.33$$

$$C = 2 \times (85 + 1.96^2) = 177.68$$

and

$$\frac{13.84 - 9.33}{177.68} = 0.0254 \quad \text{to} \quad \frac{13.84 + 9.33}{177.68} = 0.130$$

Thus the proportion is 6 % with a confidence interval, CI of 2.5 %–13 %.

NB: The CI is not symmetrical around the proportion.

The CI estimated according to the "traditional method" (Eq. 52) is

$$\pm 1.96 \times \sqrt{\frac{\frac{5}{85} \times \left(1 - \frac{5}{85}\right)}{85}} = 0.050,$$

thus from 0.009 % to 0.11 % or 0.9 % to 11 %.

At low proportions the confidence interval with this approach may produce impossible negative numbers of the probability. Try a positive diagnostic sensitivity of 3 %! (The confidence limit is negative for $p < 3.7$ and larger than 1, for $p > 96.3$.)

Uncertainty of the Difference Between Two Proportions

The general rule of error propagation is applicable:

$$u(p_1 - p_2) = \sqrt{\left(s(\overline{p}_1)\right)^2 + \left(s(\overline{p}_2)\right)^2} = \sqrt{\left(\frac{p_1 \times (1 - p_1)}{n_1}\right)^2 + \left(\frac{p_2 \times (1 - p_2)}{n_2}\right)^2} \tag{58}$$

$$z - \text{score}: z = \frac{x_i - \overline{x}}{s(x)}; \tag{59}$$

The *z-score* defines the position of an observation in relation to the average; the position is expressed in standard deviations. With this definition the z-score only makes sense if the data is normally distributed.

The 95 % confidence limit would thus be expressed as $\pm 1.96 \times z$.

The z-score is often used to normalize results from different samples and make them comparable, e.g., expressing the performance of measurements by different procedures and different quantity values.

The probability that a value is outside (larger than) the z-score is $1-NORM.S.DIST(z)$. For a two-sided evaluation use $2 \times (1-NORM.S.DIST$ $(z))$ or a normal table.

3.4   Poisson Distribution

The Poisson distribution can be understood as a special case of the binominal distribution when studying large numbers with a rare (not zero) but constant occurrence of "successes." Unlike the binominal distribution the Poisson distribution can be any number of events during the study period. The distribution describes independent and random occurrences of events in a defined time or space, e.g., radioactive decays per time unit (frequency) (dpm), blood cells in a counting chamber (cells/L), or customers in a shop during a certain time interval.

The distribution is

$$p(X) = e^{-\mu} \times \frac{\mu^x}{x!} \tag{60}$$

where μ is the observed number (radioactive decompositions, cells) usually the average observed in a unit of time or space. The probability ($p(X)$) can then be calculated for any exact value of x, i.e., the occurrence of any other number in that unit of time or space.

The distribution is defined by the value of μ, which must be equal to or larger than zero. The probability of finding the exact value is solely depending on the number of observations. The probability density function is skew to the right (it cannot be <0) but has no upper limit. With more observations the distribution approaches a Gaussian function; with 15 or higher numbers this approximations is justified.

In EXCEL, *POISSON.DIST(x,average,true)* gives the cumulative frequency and *POISSON.DIST(x,average,false)* the frequency.

Example: The average number of cells in a in a B-cell of a Bürker chamber (10^{-5} mL) is 5, what is the probability to find a B-cell containing 6 cells?

$$p(X) = e^{-5} \times \frac{5^6}{6!} ; EXCEL \ EXP(1)^{\wedge}(-5) \times 5^6/(FACT(6)) = 0.146$$

Variance of the Poisson distribution: the variance is equal to the number of observed events (average), μ.

Consequently,

Standard deviation of the Poisson distribution: $\sqrt{\mu}$ (61)

$$\text{Coefficient of variation: } \%CV\text{:}100 \times \frac{\sqrt{\mu}}{\mu} = \frac{100}{\sqrt{\mu}} \qquad (62)$$

Example: The radioactivity due to radon in a house was $340\,\text{Bq/m}^3$ (1 bequerel, Bq = 1 disintegration/s).
The standard deviation was $\sqrt{340} = 18\,\text{Bq/m}^3$ and the RSD 0.054 or % CV 5.4 %.

3.5 Set Theory and Boolean Algebra

3.5.1 Set Theory and Symbols

The theory of sets is essential for mathematics but may not immediately seem useful in day-to-day practical laboratory work. Basically, a set has some elements in common. The elements may be anything, including numerical information. Sets and their combinations are often illustrated by Venn diagrams. If, for instance in the set "all," A and B were identified, A and B might be represented by partly overlapping areas within a rectangle which represents "universe" = "all" (Fig. 5). Obvious combinations are A *and* B, A *or* B, A *not* B, and B *not* A. "A *and* B" requires the simultaneous presence of elements from both sets, "A *or* B" an element from one—or both—of the sets is sufficient. Mathematicians use special shorthand writing in dealing with sets; in the laboratory context it is useful to remember that *and* is represented by ($\cap$) meaning interaction, understood as multiplication, i.e., limiting conditions, whereas *or* are additive and recognized as union ($\cup$) of the sets A or B. Exclusions are symbolized by (\) and complements X^c. Expressing that C is a subset of A is C($\subset$) A and that the element x is included in the set A is written $x \in A$.

Example: In the universe of animals you can identify those who walk on two legs (A) and those who fly (B) as two different sets. Some animals have two legs and can also fly. Therefore, A ($\cap$) B will be limited to those who can both fly *and* walk on two legs, essentially birds (but not penguins!).

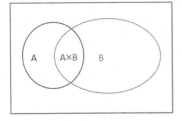

FIGURE 5 Venn diagram. The rectangle represents the universe, A and B partly overlapping datasets. $A \times B$ is the dataset that satisfies the definition of both A and B.

A^c ($\cup$) B^c will represent all other animals, e.g., snakes and fish which do not have two legs nor fly; pronounced "complement to A *or* B."

3.5.2 Boolean Algebra

Boolean algebra or logic evaluates the variables, statements, or "propositions" (A or B etc.) as "TRUE" or "FALSE," often symbolized by 1 and 0, respectively. Although looking like numbers they are not numbers, only symbols for TRUE and FALSE; a proposition can have no other values. The propositions are connected by the three basic logical connectors AND, OR, and NOT. The outcome of two propositions connected by AND requires that all propositions are TRUE which is limiting whereas if connected by OR it is enough if one of the propositions is TRUE and more alternatives would satisfy the proposition (Table 6). Compare with set theory! Besides statistical inference, Boolean logic is much applied in electronics where "switches" may be either open or closed, thus interpreted as 1 or 0 in digital computing.

The outcome of the simple rules of Boolean algebra can be summarized in a "truth table" (Table 6):

Like in algebra there are extensive and detailed rules how calculations can be performed in Boolean algebra. The rules are fundamentally different from those of conventional algebra. Referring to Venn diagrams (Fig. 5) may facilitate the understanding.

Boolean algebra is used in basic search strategies as applied to "search engines." Operators are not always explicit and then usually OR-logics are assumed, i.e., basically all mentioned propositions are considered. Special algorithms are then used to select results which satisfy conditions which make them more likely to be significant or wanted.

TABLE 6 Truth table. A and B are propositions and res the outcome if the conditions 0 or 1 are fulfilled

AND			OR			NOT	
A	B	Res	A	B	Res	A	Res
0	0	0	0	0	0	0	1
0	1	0	0	1	1	1	0
1	0	0	1	0	1		
1	1	1	1	1	1		

Thus, if in the "OR" section, third line from below A is FALSE (0) and B is TRUE (1) then the result is TRUE (1), whereas the same values of the proposition under AND and NOT conditions would be interpreted as rejection.

4. ROBUST ESTIMATORS

Robustness is a property of the algorithm, formula, or procedure used to estimate a quantity, not of the estimates it produces, e.g., results or conclusions. Deviations from assumptions (e.g., normality, homoscedasticity) may lead to erroneous results or conclusions and robust estimators are supposed to be resistant or tolerant to such deviations. This means that if the assumptions are only approximately met, the robust estimator will still have a reasonable efficiency, and result in a reasonably small bias, as well as being asymptotically unbiased, meaning having a bias tending towards zero as the sample size tends towards infinity. It is sometimes recommended that robust estimators should be used rather than trimming or truncation of datasets. However, practice varies between disciplines.

Robust estimators and nonparametric methods usually have an inferior statistical efficiency compared to parametric estimators.

4.1 Central Tendency: Median and Mode

4.1.1 Median (50 percentile)

In an ordered set of numbers, the number in the middle is the median. If the dataset consists of an odd number of observations, the median is the middle value, if the number is even it is the average of the two adjacent middle numbers. The median of a finite list of numbers can thus be found by arranging all the observations from the lowest value to the highest value and picking the middle one. Since the median is the quantity value in the middle of an ordered dataset it is unaffected by the values of the observations except the middle value; outliers do not change the value of the median as long as their order is unaffected and the number of observations is the same above and below the median. The median can thus be viewed as a robust estimate of the average.

In a series of n odd ordered numbers:

$$\text{Median} = x_{(n+1) \times 0.5} \tag{63}$$

In a series of n even ordered numbers:

$$\text{Median} = \frac{x_{(n \times 0.5)} + x_{(n \times 0.5+1)}}{2} \tag{64}$$

EXCEL offers a function to calculate the median directly from an array of values: *MEDIAN(array)*.

4.1.2 Mode

The mode of a discrete probability distribution is the value x at which its probability function takes its maximum value or peak. In other words, it is the exact value that is most likely to be sampled. A density function may have several peaks and is then referred to as multimodal in contrast to unimodal.

Clearly, situations occur where no measured values are identical and the mode is no single number. To estimate a mode under those conditions, the dataset is split into intervals and the number of observations within each interval (bin) calculated. The count of all observations is attributed to the average of the interval. This is also the way a histogram is created. Particularly if the dataset contains few observations the size, beginning, and end of the intervals will be crucial for the value of the mode.

EXCEL offers two functions to identify the mode function *MODE.MULT (array)* is suitable for multimodal distributions. The function returns a vertical array of the most frequent numbers (the function must be entered as an array function). The function *MODE.SNGL(array)* returns the single most frequent number.

In a normal distribution the mode, median, and average coincide.

4.2 Trimmed Averages

To evaluate the effect of outliers (however defined), trimmed and winsorized averages may be used and are often referred to as robust location estimates. Trimmed and winsorized averages should be used with care if the distribution is not symmetrical. The resulting distributions are not always Gaussian. On an average, trimming will underestimate the true dispersion.

$$\text{Average, trimmed: } \bar{x}_k = \frac{x_{(k+1)} + x_{(k+2)} + \cdots + x_{(n-k)}}{n - (2 \times k)} = \frac{1}{n - (2 \times k)} \sum_{i=k+1}^{n-k} x_i$$

(65)

In the dataset the k lowest and k highest values are deleted, i.e., eliminated symmetrically. The arithmetic average of the remaining data is the *trimmed average*.

EXCEL offers an interesting option *TRIMMEAN(array,fraction)* where fraction refers to a selected middle percentage of the data.

Example: See below.

The winsorized (this procedure is named after the statistician Charles P. Winsor (1895−1951)) average resembles the trimmed average but rather than discarding the highest and the lowest numbers, the removed numbers are replaced with the next higher and next lower number, respectively. A 90 % winsorization would cut the dataset at .the 5- and 95 percentiles, leaving the middle 90 % of the results for evaluation.

Average, winsorized:

$$w_k = \frac{(k+1) \times x_{(k+1)} + x_{(k+2)} + \cdots + x_{(n-k-1)} + (k+1) \times x_{(n-k)}}{n}$$

$$= \frac{k \times x_{(k+1)} + \sum_{i=k+2}^{n-k} x_i + k \times x_{(n-k)}}{n} \tag{66}$$

In a perfect Gaussian distribution the trimmed and the winsorized averages would remain unchanged but the standard deviation (dispersion) be reduced.

Examples: Rearrange 802; 854; 823; 790; 815; 840; 833; 809; 843; 821 (average 823.1; $s = 19.8$) in increasing order: 790; 802; 809; 815; 821; 823; 833; 840; 843; 854.

To calculate the trimmed average, delete 790 and 854:

$$\overline{x}_k = \frac{802 + 809 + 815 + 821 + 823 + 833 + 840 + 843}{8} = 823.25; \quad s = 14.6$$

Or directly *TRIMMEAN(array,fraction)* thus, *TRIMMEAN(array,0.2)* = 823.25, however, the standard deviation must be calculated independently from the obtained, trimmed dataset.

To calculate the winsorized average, delete 790 and add 802 (the lowest and second lowest) and delete 854 and add 843 (the highest and second highest):

$$\overline{w}_k = \frac{802 + 802 + 809 + 815 + 821 + 823 + 833 + 840 + 843 + 843}{10}$$

$$= 823.10; \quad s = 16.1$$

NB: A correct rounding would suggest $\overline{x}_k = 800 \pm 10$ and $\overline{w}_k = 820 \pm 16$ since the lowest number of significant digits is one and two for $xbar_k$ and $wbar_k$, respectively.

4.3 Dispersion of Data

There are different robust measures of dispersion and the most frequently encountered in addition to the standard deviation (variance) are the Median Absolute Deviation (MAD), the Average Absolute Deviation (AAD), and the Mean Squared Error (MSE).

Median absolute deviation, $\text{MAD} = \text{median}\left(\left|x_i - \text{median}(X)\right|\right)$ (67)

i.e., the median of the observations' differences (without sign = absolute) from the median of the observations.

Sometimes the term "robust average" is used, meaning median(X) and written $= x^*$. The asterisk indicates that the value is the result of the "robust algorithm," in this case the median.

Example: Calculate the MAD of 790; 802; 809; 815; 821; 823; 833; 840; 843; 854; median $= 822$

Calculate the absolute differences from the median:

32; 20; 13; 7; 1; 1; 11; 18; 21; 32; and rearrange in ascending order:

1, 1, 7, 11, 13, 18, 20, 21, 32, 32 and calculate the median of the differences $(13 + 18)/2$ which is the MAD $= 15.5$ (correctly rounded 16).

Outlier identification based on MAD

If $\dfrac{(\text{suspect number} - \text{median})}{\text{MAD}} > 5$ then the number is often regarded as an outlier.

Robust standard deviation from MAD (provided the data are Gaussian distributed) $s^* \approx \dfrac{\text{MAD}}{0.6745} = 1.483 \times \text{MAD}$.

This is calculated from $\dfrac{\text{MAD}}{NORM.S.INV(0.75)} = \dfrac{\text{MAD}}{0.6745}$, i.e., based on the inverse of the cumulative distribution function for the standard normal function in the third quartile (Q3).

There are other robust algorithms to estimate dispersion, particularly for small datasets.

If $n = 2$, use $\sqrt{\dfrac{(x_1 - x_2)^2}{2}} = \dfrac{x_1 - x_2}{\sqrt{2}}$.

This expression is obtained from the Dahlberg formula (33). Since it is based on only one pair of observations it has a very large and unspecified uncertainty attached.

To obtain an estimate of the standard deviation from the absolute distance to the median, use

$$s^* = \frac{\sqrt{\pi}}{n \times \sqrt{2}} \times \sum\nolimits_{i=0}^{n} \left(|x_i - \text{median}(X)|\right) = \frac{1}{0.798 \times n} \times \sum\nolimits_{i=0}^{n} \left(|x_i - \text{median}(X)|\right).$$

$$\textit{Multiple of median}: MoM = \frac{x_i}{median} \qquad (68)$$

MoM expresses the result as a fraction of the median and it is a measure of how much an individual test result deviates from the median. It can be understood as a nonparametric counterpart of the z-value, cf. z-score (Eq. 59).

The use of MoM can facilitate the interpretation and comparison of results from different methods of measurements and is therefore often used in medical screening tests. The concept was first launched in comparing results of α-fetoprotein, AFP, in screening for neural tube defects in the fetus.

$$\textit{Average absolute deviation from the average}: AAD = \frac{\sum\nolimits_{i=1}^{n} (|x_i - \bar{x}|)}{n}$$

$$(69)$$

The EXCEL function *AVEDEV(array)* calculates the AAD.
The AAD must not be confused with MAD.
The relation between these measures is MAD $<$ AAD $< s(X)$.
The MAD, AAD, and $s(X)$ have the same unit as the original data.

4.4 Percentiles, Quartiles, and Quantiles

If the median expresses the quantity value below which half of the dataset
will be found the percentile does the same for any percentage. There are two
different definitions of percentile addressing the questions

1. which is the (interpolated) value of the kth percentile? and
2. which is the (interpolated) quantity value of the data point below which
   $k\%$ of the dataset is lies? Statistical software packages may use different
   techniques for calculation of the percentile.

EXCEL offers two functions to calculate the percentile *PERCENTILE.
INC(array,k)* and *PERCENTILE.EXC(array,k)*, i.e., inclusive and exclusive.
The functions will correspond to the questions 1 and 2, respectively.
Alternative 2, i.e., *PERCENTILE.EXC(array,k)* is said to better correspond
to the definition of the percentile. In previous versions of EXCEL the only
available function corresponds to *PERCENTILE.INC(array,k)*.
The general formulas for calculating the rank (position) in an array are:

PERCENTILE.EXC:$p \times (n + 1)$
PERCENTILE.INC:$p \times n + (1 - p) = p \times (n - 1) + 1$

where n is the number of observations in the dataset and p is the percentile.
Both functions use a linear interpolation.
An easy explanation of the difference between the two functions is that
the *PERCENTILE.EXC* excludes the lowest and highest ranked numbers of
the array. This is expressed as *PERCENTILE.EXC*: $1/p < p < (1 - 1/p)$ and
PERCENTILE.INC: $0 < = p < = 1$. The manual calculation of a percentile
first requires calculating the position (rank) of the pth percentile (see above).

Example: Consider an array 2, 6, 4, 9, 11.
Calculate the 75 % by the two functions above and manually.
Calculate the rank (position, location): $0.75 \times (5 + 1) = 4.5$ and $0.75 \times$
$(5 - 1) + 1 = 4$. The percentiles calculated by the two formulas are 10 and 9,
respectively.
PERCENTILE.EXC*(array,0.75) = 10*   and   *PERCENTILE.INC
(array,0.75) = 9*.

EXCEL also offers the reverse, answering the question which percentile
is a "quantity value"?:

PERCENTRANK.INC(array,x,significance) and

PERCENTRANK.EXC(array,x,significance). "significance" refers to the number of decimals required, default is 3. **NB:** These need not be "significant."

To manually find the corresponding data it is necessary to sort the data and if necessary interpolate between results.

Example: The following procedure calculates the rank of the value nearest below the percentile and interpolates in the interval between the numbers closest to the percentile. Consider, as an example, a series of seven numbers (n), ordered and (ranked [R]): 3(1), 5(2); 7(3); 8(4); 11(5); 13(6); 20(7). Estimate the 35th (p) percentile!

$$p = \frac{100 \times R}{n + 1} \tag{70}$$

and thus $R = \dfrac{p}{100} \times (n + 1) = \dfrac{35}{100} \times 8 = 0.35 \times 8 = 2.8$. Accordingly the 35th percentile corresponds to the rank R 2.8 and the numerical value corresponding to the rank shall now be calculated:

The p percentile is obtained by interpolation and is the value corresponding to

$$\text{Integer}_R + \text{Fraction}_R \times (\text{diff.between value scored } R + 1 \text{ and } R) \tag{71}$$

In our example Integer_R is 2 and the corresponding value is 5.

The next scored value is 7 and the Fraction_R is 0.8, i.e., 2.8−2. Thus: $p_{0.35} = 5 + 0.8 \times (7 - 5) = 6.6$ which is also obtained by the *PERCENTILE. EXC* function, the *PERCENTILE.INC* is 7.1, and the median is 8.

This procedure will also give the correct median (50th percentile), i.e., $p = 0.50$:

$R = \dfrac{50}{100} \times 8 = 4$ and the 4th number in the example is 8.

Interpolation can only be made between ordered (sorted) data and between adjacently ranked data.

4.4.1 Quantiles

A general expression for dividing ordered data into q equal-sized data subsets is the q-quantiles; the quantiles are the quantity values that mark the boundaries between consecutive subsets. Quantiles are thus points taken at regular intervals from the cumulative distribution function of a random variable. Put another way, the kth q-quantile is the value x such that the probability that a random variable will be less than x is at most k/q and the probability that a random variable will be more than or equal to x is at least $(q - k)/q$.

In a dataset, there are $(q - 1)$ quantiles, with k as an integer satisfying $0 < k < q$. For instance, if you divide a length into six equal parts, there will be five boundaries.

Yet another way to describe the quantile is as a number x_p such that a proportion p of the population values is less than or equal to x_p. The quantiles are thus values which divide the distribution so that there is a given proportion of the observations below the quantile. For example, the median is a quantile, the 50th percentile.

Thus, the 0.25 quantile (also referred to as the 25th percentile) of a variable is a value (x_p) such that 25 % (p) of the values of the variable fall below or are equal to that value. Note the conceptual difference between inclusive and exclusive which was discussed above.

If a dataset is divided into 10 sections with equal number of observations, they would be recognized as "deciles."

4.4.2 Quartiles

Quartiles are the special quantiles created when the dataset is divided into four equal parts. The quartiles are thus the three limits, Q1, Q2 equal to the median, and Q4.

The first quartile, or 25th percentile (also written as Q1), is the quantity value (x_L) for which 25 % of values in the dataset are smaller than x_L.

The second quartile or 50th percentile, x_M (also written as Q2) is the median. It represents the value for which 50 % of observations are lower and 50 % are higher.

The third quartile or 75th percentile, x_H (Q3) is the quantity value such that 75 % of the observations are less than x_H.

$$\text{Interquartile interval(IQR):} p(0.75) - p(0.25) \qquad (72)$$

The interquartile interval (IQR) includes 50 % of the data and is also known as the central 50 % of the distribution.

The interquartile interval is the interval covered by the box in the standard "box-plot," i.e., from Q1 to Q3. The Q2, median, is visualized in the box. Often "whiskers" show various other parts of the distribution, e.g., the 10 % and 90 % or parts related to a normal distribution, e.g., $\pm 2s(X)$. Box plots occur in many forms, e.g., the width may be proportional to the number of observations or "notched" to show a confidence interval.

Example: Calculate the interquartile interval in EXCEL: *PERCENTILE.EXC (array;0.75) − PERCENTILE.EXC(array,0.25)*,

NB: There are the functions *QUARTILE.EXC(array,quart)* and *QUARTILE.INC (array,quart)* which split the dataset into four equal parts according to the "quart" which therefore only can take integers between 1 and 4). The IQC interval is *QUARTILE.EXC(array,3) − QUARTILE.EXC(array,1)*.

4.4.3 Quantile-Quantile plots to compare distributions

Quantiles can be used to compare the type of frequency function of two distributions, e.g., investigate the normality of an experimentally found distribution. The quantiles of an unknown distribution are calculated and compared with those of a known (e.g., normal) distribution using regression analysis. If the distribution of the data coincides with the assumed distribution, the data will follow a linear regression, if the functions are identical the regression will coincide with the equal line and there might sometimes be an intercept indicating a bias between the distributions. The results are commonly presented in a Q–Q (Quantile–Quantile) plot which thus is a graphical method to display an overall impression of one distribution in relation to another (Fig. 6). The technique of the Q-Q plot, also known as "inverse frequency plot" is not limited to comparing a data set to a normal distribution, comparisons between any distributions can be made and the same

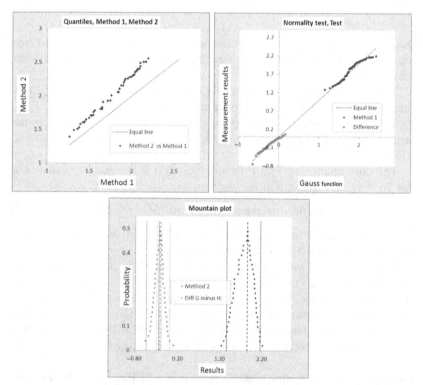

FIGURE 6 Q-Q plots. In the upper left panel the quantiles of two datasets are compared and the linear regression indicates that the frequency distributions are similar but there is a bias, a difference in location. In the middle panel the quantiles of one of the methods (high right) and those of the difference between the results (low left) are compared with a normal frequency distribution. Their close proximity to the equal line indicates that they are close to normal. In the lower panel the distribution of the data is shown as empirical cumulative plots "mountain plots" of the differences (left) and the measurement results. At least the mountain plot of the differences is almost symmetrical where as the mountain plot of the results of method 1 shows a left skewness which is indicated also in the Q–Q plot. The vertical lines indicate the medians and delineate central 95‰, respectively.

interpretations applied. Properties of a distribution which can be evaluated are primarily the location and skewness.

Quantiles are also used to estimate an empirical cumulative frequency plot, i.e., often called a mountain plot. The theory and estimation of mountain plots will be further discussed in the section on comparison of methods.

More elaborate tests for normality are the Kolgomorov–Smirnov and Anderson–Darling tests which quantitate the divergence from normality. Essentially these methods quantify the differences between the normal density function and each of the underlying columns in a corresponding histogram with defined bin widths. These methods have their advantages and disadvantages; generally the Kolgomorov–Smirnov method is said to be less powerful than the other. The Kolgomorov–Smirnov formula can also be used to estimate the "goodness of fit."

4.4.4 Probit

The probit (probability unit) function is the inverse cumulative distribution function of the normal distribution. The probit function thus generates a value of a variable, associated with a specified cumulative probability. The probit only exists for values between 0 and 1.

In EXCEL the function is calculated by *NORM.S.INV(probability)* and can be used to simulate the sigmoid curve for probabilities between 0 and 1 (see Fig. 2).

4.4.5 Logit

If probit is the inverse function of the probability, then the logit is the function of the logarithm of the probability, expressed as odds (see Eq. 48)

$$\text{logit}(p) = \ln\left(\frac{p}{1-p}\right) = \ln(\text{odds}) = \ln(p) - \ln(1-p) \tag{73}$$

where p is the probability, see Eq. (194). The choice of base for the logarithm is arbitrary but the natural base e is commonly used.

4.4.6 Odds Ratio

The ratio between two odds (O_1 and O_2) is recognized as the odds ratio (OR) (Eq. 224)

$$\text{OR} = \frac{O_1}{O_2} = \frac{\dfrac{p_1}{1-p_1}}{\dfrac{p_2}{1-p_2}} = \frac{\dfrac{p_1}{q_1}}{\dfrac{p_2}{q_2}} = \frac{p_1 \times q_2}{p_2 \times q_1} \tag{74}$$

Thus, the difference between two logit values is the logarithm of the odds ratio

$$\ln(\text{OR}) = \ln\left(\frac{\frac{p_1}{1-p_1}}{\frac{p_2}{1-p_2}}\right) = \ln\left(\frac{p_1}{1-p_1}\right) - \ln\left(\frac{p_2}{1-p_2}\right) = \text{logit}(p_1) - \text{logit}(p_2)$$

(75)

5. UNCERTAINTY

Error may perhaps be a nebulous concept, e.g., by definition analysts do not make errors. Nevertheless all results suffer an attached "uncertainty" which can be attributed to random and/or systematic variation, sometimes synonymous with error. Properties and methods to estimate the uncertainty in measurements have been summarized as the "uncertainty concept," MU. The internationally normative standard for estimating measurement uncertainty is "ISO-BIPM-JCGM 100: Evaluation of measurement data—Guide to the expression of uncertainty in measurements" known as "GUM" and freely available on the internet.

The uncertainty is expressed as the standard uncertainty of a process and attached to the results of measurements. The principle of the uncertainty concept is that all error-prone steps in a measuring system, which may affect the result, shall be identified, quantified, and presented in an uncertainty budget. The uncertainty of an identified item is referred to as the "standard uncertainty," u. Since the standard uncertainty is expressed as, and equal to, the standard deviation, it is the same "kind-of-quantity" and expressed in the same unit as the measurand.

The uncertainty should be understood as an interval within which the true, conventionally true, or reference value is supposed to be found with a defined level of confidence. The uncertainty being attached to a measurement result is equal to the inaccuracy where accuracy is defined and "closeness of agreement between a measured quantity value and a true quantity value of a measurand." It thus refers to an individual result which has been corrected-as far as is possible-for a systematic error.

5.1 Uncertainty Budget

The basis for the uncertainty estimate is an uncertainty budget, which lists all identified and quantified sources of uncertainty of the measurement procedure. Any identified and significant bias shall be eliminated or corrected and, if necessary and appropriate, substituted by an uncertainty that shall be included in the uncertainty budget. Eliminating the bias will increase the uncertainty of the result but the reported, best estimate has a potency to come closer to the "true" value of the measurand. Most importantly, as a result, the laboratory takes the responsibility for containing a bias which the user normally cannot do.

The quantities in the uncertainty budget shall be combined according to the measurement function which describes their interactions. This can be cumbersome if the components are many and the functional relationships complex. In essence the Gaussian quadrature propagation rules are used (Eqs. 77−80). This results in the "combined uncertainty" u_c.

The combined uncertainty is comparable to the standard deviation of the procedure and requires adjusting to reach the desired probability. The level of confidence of the estimated uncertainty may be adjusted by multiplying the combined uncertainty by a "coverage factor," k, to obtain the

$$\text{Expanded uncertainty}(U)\colon \quad U(X) = k \times u_c(X) \tag{76}$$

The value of k shall always be included in a report if a result is given with an expanded uncertainty attached. Without this information the uncertainty is regarded as covering only about 65 %, i.e., $k = 1$.

The level of confidence does not have the same stringency as "confidence interval" which also considers the type of distribution (e.g., normal or Student's distribution). In presenting the uncertainty interval, a k-value of 2 is usually considered to create a level of confidence of 95 %, likewise, a $k = 3$ results in a level of confidence of 99 %.

NB: The relative uncertainty is always calculated from the combined standard uncertainty, not the expanded uncertainty (cf. the Note to Eq. (29)).

5.2 Propagation of Measurement Uncertainty

5.2.1 Propagation Rules

The combined uncertainty u_c of additions and subtractions:

$$X = x_1 \pm x_2 \pm \cdots \pm x_i \tag{77}$$

is

$$u_c(X) = \pm \sqrt{u(x_1)^2 + u(x_2)^2 + \cdots + u(x_i)^2} \tag{78}$$

In general terms, variances can be added, however, compare weighting the variances in pooling variances (Eq. 30).

Example: The variables $A = 10$ and $B = 21$ have the standard uncertainties of $u(A) = 0.3$ and $u(B) = 0.6$; the sum a combined uncertainty $u_c(A + B; \; 31) = \pm \sqrt{0.3^2 + 0.6^2} = 0.67$.

The combined uncertainty u_c of multiplications and divisions:

$$X = x_i \times (\div) x_2 \times (\div) \times \cdots \times (\div) x_i \tag{79}$$

is

$$\frac{u_c(X)}{X} = \pm \sqrt{\left(\frac{u(x_1)}{x_1}\right)^2 + \left(\frac{u(x_2)}{x_2}\right)^2 + \cdots + \left(\frac{u(x_i)}{x_i}\right)^2} \tag{80}$$

Example: The ratio between the variables (10 ± 0.3 and 21 ± 0.6) is $\dfrac{10}{21} = 0.48$. Estimate the combined relative and absolute uncertainty of the ratio!

$$u_{c\text{-rel}} = \frac{u_c\left(\dfrac{A}{B}\right)}{\dfrac{A}{B}} = \frac{u_c\left(\dfrac{10}{21}\right)}{\dfrac{10}{21}} = \sqrt{\left(\frac{0.3}{10}\right)^2 + \left(\frac{0.6}{21}\right)^2} = 0.041$$

and the absolute uncertainty thus amounts to $u_c(0.48) = 0.048 \times 0.041 = 0.020$.

If the variables were multiplied, the relative uncertainty would be the same (0.041) but the absolute different since the product is 210. Thus $u_c(210) = 8.70$.

For simplicity the uncertainties or relative uncertainties, as appropriate, may be added linearly. This always leads to an overestimation of the combined uncertainty, the magnitude of which depends on the relation between the uncertainties of the terms and factors. In the examples above the estimations would be 0.90 instead of 0.67 and 0.058 instead of 0.041, respectively.

A more comprehensive method to estimate the uncertainty of a function $q(x,\ldots,z)$ involves the use of partial derivatives

$$u_c(q) = \sqrt{\left(\frac{\partial q}{\partial x} u(x)\right)^2 + \cdots + \left(\frac{\partial q}{\partial z} u(z)\right)^2} \tag{81}$$

In case the variables (e.g., $x,\ldots,z$) are not independent the covariance between all the variables must be taken into consideration

$$u_c(q) = \sqrt{\left(\frac{\partial q}{\partial x} u(x)\right)^2 + \cdots + \left(\frac{\partial q}{\partial z} u(z)\right)^2 + 2 \times \frac{\partial q}{\partial x} \times \cdots \times \frac{\partial q}{\partial z} \times \text{covariance}} \tag{82}$$

For estimation and definition of the covariance, see Eq. (174).

There are several shortcuts and algebraic approximations of the partial derivation. A frequently used method that is applicable to spreadsheet programs is that of Kragten (ref. see Eurachem Guide CG4, p. 104). This requires that the variables are uncorrelated. The approximation handles complex formulas, e.g., exponentials, inadequately. When the same quantity occurs more than once in a measurement function the Kragten approximation may overestimate the uncertainty due to so-called compensating errors.

The uncertainty of a result can also be estimated by Monte Carlo simulation of input variables and combining them according to the measuring function. This requires that the frequency distribution of the individual input variables and their uncertainty can be adequately estimated. In most cases it is reasonable to assume that measurement results are normally distributed but some input variables may be better represented by a rectangular or triangular distribution.

5.3 Estimates of Uncertainty

GUM, the international standard describing measurement uncertainty, rules that a combined uncertainty shall be estimated after dissection of the measurement procedure, identifying the input variables and their individual uncertainties and combining them as described above. This is known as the "bottom-up" method. However, sometimes this approach is not feasible and therefore GUM opts for a "top-down" procedure by which a combined uncertainty is achieved by repeated measurements of the quantity.

If the bottom-up procedure is chosen the standard uncertainty should be calculated by statistical methods as a first priority (Type A). If not possible or feasible, the standard uncertainty may be estimated by other methods, e.g., experience, literature, i.e., "scientific judgement" (Type B). The Type A is usually equivalent to an evaluation of repeated measurements, assuming a normal distribution of the results. The rectangular and triangular distributions are frequently used to aid estimates of uncertainty according to Type B. Estimates by Type A and Type B are treated equally in an uncertainty budget.

5.4 Rectangular and Triangular Distributions

5.4.1 Standard Uncertainty of a Rectangular Distribution

If all results were distributed within an interval $(2a)$ between an upper (UL) and lower limit (LL) and the probability for a specific value were the same in the entire interval, then the distribution would be recognized as "rectangular" or "uniform." Consequently, the "top" of the frequency distribution will be horizontal and the cumulative frequency distribution rectangular. Extreme values, i.e., those close to the upper or lower limit of the distribution, will be as probable as any other within the distribution. No values would be possible outside the assumed interval.

The uncertainty estimated for a rectangular distribution is the most conservative, i.e., gives the largest "standard uncertainty."

$$u(X) = \pm \frac{a}{\sqrt{3}} \tag{83}$$

where $2a$ is (*Upper Limit − Lower Limit*) and equal to the AUC.

Example: Assume that the length of a rod is no less than 50 cm and no longer than 150 cm and estimate the length and uncertainty! The best estimate of the length is 100 cm, i.e., the middle of the interval and the uncertainty estimated

$$u(\text{rod}) = \frac{150 - 50}{2 \times \sqrt{3}} = \pm 28.9 \text{ cm}.$$

NB: The standard deviation according to Eq. (83) represents about 29 % of the total distribution ($2a$). It should be compared to the standard deviation of the normal distribution which represents about 34 % of the distribution. The background is that the standard deviation is defined as the square root of the variance which is defined as the difference between two expectancies $(\text{Var}(X) = E(X)^2 - (E(X))^2)$, cf. Eq. (18).

5.4.2 Standard Uncertainty of a Triangular Distribution

If one value is more likely than other values within an interval and no values possible outside the interval, then a triangular distribution may be proper. Extreme values close to the limits of the assumed interval are possible but less likely than elsewhere in a symmetrical triangular distribution:

$$u(X) = \pm \frac{a}{\sqrt{6}} \tag{84}$$

The triangular distribution characterized by a lower and upper limit ($\text{LL} = b$ and $\text{UL} = c$) and a mode d. The average of the distribution is then

$$\mu = \frac{b + c + d}{3} \tag{85}$$

The variance is

$$\sigma^2 = \frac{(b-c)^2 + (b-d)^2 + (c-d)^2}{36} = \frac{b^2 + c^2 + d^2 - b \times c - b \times d - c \times d}{18} \tag{86}$$

In the case of a symmetrical triangular distribution, i.e., $d = \frac{b+c}{2}$, the square root of Eq. (86) is equal to Eq. (84), whereas in a "right-angle triangular distribution," i.e., $b = d$, it is simplified to

$$\sigma^2 = \frac{(b-c)^2}{18} = \frac{(2a)^2}{18} \tag{87}$$

The square root of the variance is the standard uncertainty, cf. Eq. (86)!

$$u(X) = \frac{(b-c)}{3 \times \sqrt{2}} = \frac{2a}{3 \times \sqrt{2}} \tag{88}$$

Since the expected value cannot be $<b$ or $>c$ the uncertainty is "one-sided," i.e., $b - u$ or $c + u$.

The standard deviation calculated accordingly will comprise about 41 % of the distribution.

5.4.3 Estimation of the Standard Uncertainty of a Gaussian Distribution

When the measurement value is most likely to be near the center of an interval but there is a small, but real, possibility that there might be values outside the assumed or observed limits ($\bar{x} \pm a$) then the appropriate "density function," i.e., distribution, is often assumed to be Gaussian. Since the individual values are not known the standard deviation cannot be calculated by formula (15 or 18) but estimated by

$$u(X) = \pm \frac{a}{\sqrt{9}} \qquad (89)$$

where $2 \times a$ is (Upper Limit − Lower Limit).

NB: Obviously, the estimated $u(X)$ depends on the estimate of the half-width a and the estimated standard deviation will comprise about 33 % of the distribution.

5.5 Reference Change Value and Minimal Difference

A common problem for a user of results of measurements is to estimate if results are different or different from a reference (control) value, given their uncertainty. This can be approached in terms of error propagation; in a clinical setting known as "Minimum difference" or MD.

Example: If the results of two measurements are A and B with the uncertainties $u(A)$ and $u(B)$, a significant difference (D) must be larger than the uncertainty of the difference $u(D)$ at the given level of confidence.

Therefore, the reference to 1.78 in the brackets after the formula shall be a "link",

$$D > u(D) = \sqrt{u(A)^2 + u(B)^2}, \quad (\text{cf. } 1.78) \qquad (90)$$

Example: The desirable level of confidence is usually chosen to about 95 %, i.e., a k-value of 2 or 1.96 for a normally distributed dataset. It may be reasonable to assume that $u(A) = u(B)$ and thus the minimum difference (MD) is

$$\text{MD} \geq 2 \times \sqrt{2 \times u(A)^2} = 2 \times u(A) \times \sqrt{2} = 2.77 \times u(A)$$

As a rule of thumb the MD is often given as $\text{MD} = 3 \times u(A)$.

In clinical chemistry the "Reference change value," RCV, also includes the biological variation. The variation is usually expressed in relative terms (% CV(A)):

$$RCV> = \sqrt{2\times\%CV(A_a)^2 + 2\times\%CV(A_w)^2} = \sqrt{\%CV(A_a)^2 + \%CV(A_w)^2}\times\sqrt{2}$$

$$(91)$$

Where % $CV(A_a)$ is the coefficient of variation of the measurement procedure (analytical or measurement uncertainty) and % $CV(A_w)$ is the biological within individual variation.

NB: When a result is compared to a reference value—which by definition is known without uncertainty—(Eq. 90) is reduced to $D \geq k \times u(D) = k \times \sqrt{u(A)^2} = k \times u(A)$. Consequently, diagnosis can be performed with a higher confidence than monitoring, assuming equal MU.

5.6   Index of Individuality

The usefulness of reference values in clinical diagnosis is often expressed as the "Individuality Index" (*II*):

$$II = \frac{\%CV(A_w)}{\%CV(A_b)} \tag{92}$$

Where % $CV(A_b)$ and % $CV(A_w)$ are the between individual and within individual variation, respectively. If the within individual variation is large in relation to the between variation and thus the *II* high this indicates a high utility of population-based reference intervals, otherwise reference values are of less value.

Common set-points are $II \leq 0.6$ and $II \geq 1.4$ as limits for low and high utility of the *II*, respectively.

Example: The *II* for S-Creatinine concentration is reported to about 0.3 and for S-Iron concentration about 1.1. Consequently a population-based reference interval for S-Creatinine is less reliable than that of S-Iron.

5.7   The Rule of Three

This rule states that if no event happens in *n* observations, then it can be assumed, with a probability of 95 %, that it will happen less frequently than 1 in *n*/3 observations. The factor 3 is derived from an approximation of the binominal distribution resulting in $-\ln(0.05) = 2.996 \sim 3$. This rule is an acceptable approximation if $n > 30$.

Example: If in a series of measurements there were no outliers in 600 consecutive measurements, it can be concluded with a 95 % confidence that there will be less than 1 outlier in 200 measurements, i.e., less than 0.5 %.

5.8 Monte Carlo Simulation

In a Monte Carlo simulation values are randomly drawn from a defined distribution. It is anticipated that these values are representative for the distribution. Samples may be drawn from a normal distribution or from rectangular, triangular, or other distributions.

To apply the MC simulation to uncertainty estimates, the quantities specified in the uncertainty budget are randomly and independently varied, i.e., drawn from the assumed and simulated distribution. The simulation is usually based on a standard uncertainty (dispersion) of the individual quantity and the location (value) chosen to be representative of what is expected in the actual measurement. By combining the simulated quantities according to the measurement function the same number of simulated results is obtained. The combined uncertainty can then be estimated by standard statistical methods. This procedure should be iterated at least 10,000 times but ten-or a hundred times more iterations are not uncommon.

Simulating normal distributions may be accomplished using EXCEL, e.g., the function average$(X) + s(X) \times NORM.S.INV(RAND())$ where average (X) determines the location and $s(X)$ the dispersion of the dataset. The function $RAND()$ returns values between 0 and 1 from a rectangular distribution and can thus be taken as the probability for the occurrence of a particular value.

There are other strategies to simulate normal functions in EXCEL and there is in innate function available.

Data from a triangular distribution is slightly more complicated if a skew triangular distribution should also be considered.

For simple measurement functions simulations to estimate the uncertainty may be an "overkill" and Type A or Type B estimates preferred.

Example: An uncertainty budget lists quantity A as $10 \pm u(A)$ 0.3 and B as $21 \pm u(B)$ 0.6. It is reasonable that the results of the measurements are normally distributed. Estimate $u_c(A + B)$ and $u_c(A \times B)$ by propagation rules and the average of 25 sets of 10,000 simulations.

	Propagation	Simulation
$u_c(A + B)$	$\sqrt{0.3^2 + 0.6^2} = 0.671$	0.669
$u_c(A \times B)$	$210 \times \sqrt{\left(\frac{0.3}{10}\right)^2 + \left(\frac{0.6}{21}\right)^2} = 8.700$	8.710

The number of decimals in the table was chosen for demonstration purposes. Slightly different results may be obtained on repeated simulations

5.9 Bootstrapping

Bootstrapping refers to a test that relies on "random sampling with replacement." This can be illustrated as a basket with fruits of various kinds. Blindfolded you pick a fruit, register its features, and replace it. Then take another, register, and replace it and so on as long as desired. Obviously it is possible, even likely, that the same fruit is taken more than once.

Suppose that the basket contains numbers rather than fruits. A set of numbers, the "seed," is created or obtained from an experiment. By repeating copying of a random number from the basket into a new matrix, a representative new dataset is formed which is supposed to resemble the population from which the initial sample was drawn. Consequently, a new average and variance can be calculated to form a new dataset which is assumed to better represent the population. The large number of observations will move the distribution of the data towards a normal distribution and allow the use of parametric statistical methods.

This can be achieved in EXCEL using the *INDEX* function:

$$INDEX(array, INT(ROWS(array) \times RAND() + 1), INT(COLUMNS(array)$$
$$\times RAND() + 1)),$$

which essentially selects a random number in a two-dimensional array without "consuming" it.

Bootstrapping is used to improve the accuracy (trueness and precision) of the average and variance without increasing the need for a larger number of individuals in the sample. In laboratory medicine reference intervals are frequently estimated as the average value of a defined sample population (reference population) ± 2 standard deviations. Conventionally the reference population is recommended to comprise at least 125 individuals to achieve a reliable reference interval. Applying bootstrapping would allow fewer reference individuals with a potency to achieve the same reliability of the reference interval.

We have only one dataset. When we compute a statistic from the data, we only know that dataset, which may not be a true representative sample of the population; e.g., we don't really know how variable the data might be. The bootstrap technique creates a large number of new datasets but does not create any new data, only different distributions. If the seed does not comprise enough observations the bootstrapping may have a tendency to underestimate the variance of the population. This effect is less pronounced for the central tendency (averaged and median) than for the variance of the bootstrapped data.

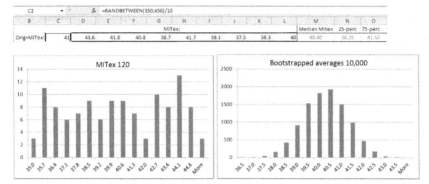

FIGURE 7 Screendump from EXCEL illustrating the distribution in of the "seed," i.e., the resulting rectangular distribution and the distribution of the averages of the bootstrapped distributions after 10,000 iterations.

Example: A sample dataset, the seed, comprised 120 randomly selected values between 35 and 45 units and labeled "MITex." The numbers were generated by the pseudorandom generator *(RAND())* in EXCEL and anticipated to represent a rectangular distribution. The standard deviation was calculated to 2.89 units (Eq. 83).

A bootstrapped "population" of 10,000 random samples was conjured from the seed by *INDEX(MITex,INT(ROWS(MITex) × RAND() + 1),INT (COLUMNS(MITex) × RAND() + 1))*. This formula was entered in each of the 10,000 cells creating a random choice from the seed, new values can be drawn from the MITex in the next cell and so on. This will give 10,000 new datasets with the same number of observations as the seed. The average and standard deviation calculated from the repeated datasets were 40.1 and 2.82. Note the "normalization" of the distributed values in Fig. 7, right panel.

Note that the number of data points in a bootstrap resample is equal to the number of data in the seed. If the average of each resampling was computed μ_i^* and this process repeated the next computed bootstrap average would be μ_{i+1}^*. If we repeat this 100 times, we have $\mu_1^*, \mu_2^*, \ldots, \mu_{100}^*$. This represents an empirical bootstrap distribution (values identified by an asterisk [*]). From this the total average and standard deviation are calculated.

6. DIFFERENCE BETWEEN RESULTS

6.1 The *t*-Distribution

If we were to decide if a value, $\bar{x}$, i.e., the average of a sample drawn from a normally distributed dataset, were different from the average (μ_0) and standard deviation (σ_0) of the known normal distribution we could express the difference as a z-value $z = \dfrac{\bar{x} - \mu_0}{\sigma_0}$. This value can be directly translated to a probability of the difference being significant with a 95 % probability, e.g., if is it larger than $\pm 1.96 \times \sigma_0$.

If the population standard deviation σ_0 is not known, i.e., we only know the sample standard deviation $(s(x))$ estimated from the sample, and the number of observations is small (<30) the comparison above is done "backwards." $\dfrac{\bar{x} - \mu_0}{s_x/\sqrt{n}}$ and the statistic "normalized deviation" (Eq. 21), known as t, follows the t distribution with $n - 1$ degrees of freedom. (See Eq. (44) where the t-distribution is discussed.) The t-distribution approaches the normal distribution when the number of observations increases. The difference between the distributions is almost negligible at $n = 50$. There are separate tables for the t-distribution to estimate the probability and AUC. There are also appropriate functions in EXCEL *T.DIST.2T*, *T.DIST.RT*, and *T.INV.2T*.

$$|t| = \frac{\bar{x} - \mu_0}{s_x/\sqrt{n}} = \frac{\bar{x} - \mu_0}{s(\bar{x})} = \frac{\bar{x} - \mu_0}{\text{SEM}}. \tag{93}$$

This is known as the single sample t-test and is evaluated using $df = n - 1$. It is applicable in evaluating a comparison between an estimated value and a known value. Also see Student's independent t-value.

Example: Suppose you have made 15 measurements of the concentration of component A in a reference material with the nominal value 10.5. Your results were 9.2, 11.0, 9.9, 10.3, 8.6, 9.2, 8.6, 9.8, 9.0, 11.3, 10.1, 9.9, 11.5, 9.9, 10.9. The "null hypothesis" is that there is no difference between the value and the average. We test if the null hypothesis is true with a given probability.

The average is 9.9 and the standard deviation 0.86.

Calculate $t = \dfrac{\bar{x} - \mu_0}{s_x/\sqrt{n}} = \dfrac{9.9 - 10.5}{0.86/\sqrt{15}} = 2.54$. The critical t-value is T.INV.2T(0.05,14) = 2.14, thus we conclude that the null hypothesis is not satisfied and therefore the difference between the value of the average and the experimental value is significant at the 95 % level.

6.2 Difference—Two Scenarios

Two different scenarios or experimental designs are identified in estimating the significance of a difference between results:

1. the difference between the averages of two different datasets (e.g., the difference in the concentration of a quantity in a group of men and a group of women), this is the "Student's independent t-value," t_{ind}; and

2. the difference between paired results (e.g., the effect of a treatment on the same individual or sample, i.e., before and after treatment, or a concentration of the same samples measured by two methods), this is the "Student's dependent t-value," t_{dep}.

Both can be resolved and evaluated using Student's t-test; it is only the variables and the formulation of the normalized deviate that differs.

6.3 Difference Between Averages; Student's t_{ind}

It is intuitive that the importance—or significance—of a difference between two results will be determined by two variables, the difference between the central tendency of the distribution of the obtained results and how well they have been determined, i.e. the uncertainty of the average. If the distribution of both datasets are normal, the difference between the averages shall be normalized by the combined standard error of the means (SEM).

This applies for instance to a comparison between results of different measurement procedures using the same set of test materials. Other examples are comparing the averages of a particular quantity among groups of people e.g., the weight of men and women a defined age interval. The observations *per se* can be performed apart from each other, e.g., in different laboratories and there may be different numbers of observations underlying the averages and their uncertainty. The measurements are thus independent of each other and the method of choice to evaluate the difference is the independent Student's t-test; t_{ind}.

The t-value can be calculated by normalizing the difference by the combined standard error of the means, according to Welch:

$$|t_{ind}| = \frac{\bar{x}_1 - \bar{x}_2}{\sqrt{(s_1^2/n_1) + (s_2^2/n_2)}} = \frac{\bar{x}_1 - \bar{x}_2}{\sqrt{s(\bar{x}_1)^2 + s(\bar{x}_2)^2}} \qquad (94)$$

where $s(\bar{x})$ is the standard error of the mean, i.e., $s(\bar{x}) = \text{SEM}$.

If $s_1 \sim s_2$ then the degrees of freedom (df) for t_{ind} is:

$$df = n_1 + n_2 - 2 \qquad (95)$$

and the formula can be simplified

$$t_{ind} = \frac{\bar{x}_1 - \bar{x}_2}{s \times \sqrt{(1/n_1) + (1/n_2)}} \qquad (96)$$

Statistical software packages calculate the t_{ind} after input of the original observations. The individual observations need not be available to calculate the t_{ind} since the only pieces of information that are necessary are the averages $\bar{x}_1$ and $\bar{x}_2$, the standard deviations s_1 and s_2, and the number of observations, n_1 and n_2 of each series of measurements—and formula (94).

If $s_1 \neq s_2$ the Welch-Satterthwaite approximation shall be used to estimate the degrees of freedom (df).

$$df = \frac{\left[(s_1^2/n_1) + (s_2^2/n_2)\right]^2}{\dfrac{(s_1^2/n_1)^2}{n_1 - 1} + \dfrac{(s_2^2/n_2)^2}{n_2 - 1}} = \frac{\left[(s_1^2/n_1) + (s_2^2/n_2)\right]^2}{\dfrac{s_1^4}{n_1^2(n_1 - 1)} + \dfrac{s_2^4}{n_2^2(n_2 - 1)}} = \frac{\left(se_1^2 + se_2^2\right)^2}{\dfrac{se_1^4}{(n_1 - 1)} + \dfrac{se_2^4}{(n_2 - 1)}}$$

(97)

A df estimated by this procedure is not necessary an integer and needs to be rounded if used as an entry to a t-table. The use of the Welch-Satterthwaite approximation gives a conservative estimate of the significance.

The significance of a difference between distribution variances (standard deviations) and thus the need to apply the Welch-Satterthwaite approximation can be estimated applying an F-test (Eq. 107). It should be observed that the approximation does not affect the t-value, only its interpretation caused by the change of df.

EXCEL supports two procedures for calculating the t_{ind}; when equal or unequal variances are assumed "Two Sample Assuming Equal Variances" and "Two Sample Assuming Unequal Variances," respectively. This requires that an "Add-in" is activated, which occurs under "Data." It is safe to use the latter option since the results will coincide if the variances are not different. The reports comprise the descriptive statistics of the groups, the calculated t-values and the critical values and corresponding probabilities for one- and two-tail evaluations.

For a manual solution, the probabilities can be calculated from the t-values by using the EXCEL functions $T.DIST.2T(ABS(t\text{-}value),df)$ and $T.DIST.RT(ABS(t\text{-}value),df)$, for a two- and one-tailed test, respectively, and the critical t-values calculated using $T.INV(probability,df)$ or $T.INV.2T$ $(probability,df)$ for one- and two-tail evaluations, respectively.

The null hypothesis of the Student's independent t-test assumes that there is no difference between the averages. Thus, a t-value that exceeds the critical value indicates a significant difference with a probability calculated with any of the $T.DIST$ functions.

The uncertainty of the quantity value of a reference material is usually expressed as a confidence interval or as the standard error, i.e., how well the value has been determined. Assuming the $s(\bar{x})$ of the reference material to be $s(\bar{x}_{RM})$ the formula will be

$$t_{ind} = \frac{\bar{x}_1 - \bar{x}_{RM}}{\sqrt{\dfrac{s_1^2}{n_1} + s(\bar{x}_{RM})^2}}$$

(98)

The quantity value of the reference material may be an assigned value without an uncertainty attached. Then Eq. (98) can be further simplified and the $s(\bar{x}_{RM})^2$ disregarded.

The Student's independent t-test assumes a Gaussian distribution of the datasets. If this is not the case the nonparametric method of choice is *Mann-Whitney U-test* which is described in some detail below.

6.4 Difference Between Paired Results; Students t_{dep}

When results are obtained from measuring the *same* sample with different methods or samples from the *same* individual at different points in time, the difference between the individual results can be used to estimate the difference between procedures. The same individual, or sample, is used twice the results are not considered independent. Since each difference is considered this approach may intuitively be more sensitive, however this concept is defined.

In this case the t-value is calculated by normalizing the average of the differences, i.e., by the standard error of the distribution of the differences. Therefore, if n pairs were measured, the difference between the results was d_i, the average of the differences between the pairs $\bar{d}$, and the standard deviation of the differences s_d, then

$$\left|t_{\text{dep}}\right| = \frac{\bar{d}}{s_d/\sqrt{n}} = \frac{\sum_{i=1}^{i=n} d_i/n}{s_d/\sqrt{n}} = \frac{\left(\sum_{i=1}^{i=n} d_i/n\right)\sqrt{n}}{s_d} = \frac{\sum_{i=1}^{i=n} d_i}{s_d \times \sqrt{n}} \qquad (99)$$

Since the Student's t_{dep}-*test* is based on an estimated average and standard deviation, it assumes a Gaussian distribution of the differences between the observations. The distribution of the individual datasets is not important.

$$\text{The degrees of freedom for } t_{\text{dep}}\text{: } df = n - 1 \qquad (100)$$

EXCEL supports calculation of the t_{dep}. This requires that an "Add-in" is activated. It is available under Data and called "Paired Two Sample for Average." The table of results which is calculated includes descriptive statistics of the groups, calculated t-*values* for one- and two-sided problems and the corresponding critical t-values and probabilities.

There is a command *T.TEST(array1,array2,tails,type)* which directly interprets the t-test in terms of the probability. This command is applicable for t_{dep} as well as t_{ind} (see below). The probability is readily obtained *T.DIST.2T(ABS(t-value),df)* or *T.DIST.RT(ABS(t-value),df)* for two- and one-tail situations, respectively.

Critical t-values are calculated by the functions *T.INV(probability,df)* or *T.INV.2T(probability,df)* for one- and two-tail evaluations, respectively.

The null hypothesis H_0 is that there is no difference between the values of the two series. The inference of the calculated t_{dep} is judged by comparing with (a table comprising) critical values for the current df. If the calculated *t-value* is larger than the critical *t*-value corresponding to the *df*, then the null hypothesis is rejected in favor of the opposite. The table lists, and the critical *t*-values are calculated from positive *t*-values. Since the average of the differences may be negative the absolute value of the calculated *t* shall be applied.

Like for the t_{ind} the t_{dep} can easily be calculated according to the above formula (99) provided the input data were available, i.e., information on the distribution of the differences (average, standard deviation, and the number of pairs). Although many reports in the scientific literature include Student's dependent, enough information for an external calculation and assessment is usually not available.

The evaluation of tests and the use of tables are further exemplified below.

If the differences are not Gaussian distributed the nonparametric method of choice is *Wilcoxon signed rank test.*

6.5 Comparison Between Many Series

If many independent series are compared with successive Student's *t*-tests, then the overall probability overestimates that there is a difference although there is none. This is because the probability of an unexpected result due to chance increases. One way to compensate for this is the Bonferroni correction, which rules that the α-level which each *t*-value shall be tested against is the desired α-level divided by the number of datasets to compare, e.g., if three datasets were to be evaluated at $\alpha = 0.05$ then each should be tested for $\alpha = 0.050/3 = 0.017$. This is the value which should be used for each of the comparisons and the degrees of freedom (*df*) is $n - 1$ or $n - 2$ as used for the *t*-test. EXCEL:*T.INV.2T(α;df)* for a two-sided comparison and *T.INV(α; df)* for a one-sided comparison.

In essence the Bonferroni correction makes it more difficult to achieve and eventually exceed the critical *t*-value.

The Bonferroni correction is regarded as a conservative test; in other words it becomes "more difficult" than perhaps justified to demonstrate significance of many differences between series; more results may be attributed to chance alone. Particularly, this is the case when the number of series of comparisons and their associated *p*-values increases, e.g., in gene research, and the conservatism of the Bonferroni correction may be too prohibitive. A frequently used correction is by Benjamin and Hochberg who define the "false detection rate" (FDR). The procedure is based on ranking (*i*) the obtained *p*-values and adjusting for the preset desirable detection rate (*Q*) and number of independent series (*p*-values) *n*. An adjusted *p*-value is calculated as $i \times (Q/n)$. Thus, the adjusted *p*-values can be viewed as the rank multiplied by a constant which basically is the number of series. The desirable detection rate is expressed as a fraction. The adjusted *p*-value is evaluated against the desired probability, e.g., 95 %, i.e., <0.05.

TABLE 7 Benjamini-Hochberg adjusted *p*-Values of multiple comparisons

	Series ID	P-values	Rank	P(B-H)	B-H Sign	Bonferroni
1	A	0.384	6	0.109		
2	B	0.975	10	0.182		
3	C	0.001	1	0.018	Sign	Sign
4	D	0.041	2	0.036	Sign	
5	E	0.212	4	0.073		
6	F	0.762	8	0.145		
7	G	0.986	11	0.200		
8	H	0.074	3	0.055		
9	I	0.594	7	0.127		
10	J	0.940	9	0.164		
11	K	0.340	5	0.091		
12						

The desirable detection rate was 0.20 and the number of series 12. A Bonferroni adjustment is included as comparison.

Example: The influence of 12 chemical contaminants on measurements was tested and the effect was evaluated in relation to controls. Thus there were 12 series of measurements of controls and contaminated samples and each series resulted in a significance (*p*-value) of the difference obtained by Student's *t*-test. The null hypothesis was that there would be no effect (Table 7).

6.6 Interpretation of a *t*-Value

Estimated *t*-values are interpreted using a *t*-table. The same table is used for evaluations of all *t*-values irrespective of how they have been calculated. The entries to the table are the *df* and the desirable probability.

For a given *df* find the value below, and as close to the estimated *t*-value as possible, in the table. The column in which it is found represents the probability. An estimated *t*-value can lead to different interpretations depending on the number of observations. The higher the *t*-value the less probable it is that the compared quantities are the same (the null hypothesis true).

Example: A *t*-value of 2.2 was obtained in a study. The *df* was 20.

The null hypothesis states that there is no difference between the results. In this case the probability (*p*-value) that the null hypothesis was true was <0.05 (the obtained *t*-value $>$ table value). With a "significance level" of 0.95 ($\alpha = 0.05$) the difference is judged significant on this level.

The t-value can be estimated using the EXCEL function *T.INV.2T (0.05,20)* $= 2.086$ where 2.086 is the critical t-value above which the alternative hypothesis is true, i.e., there is a 95 % probability that the difference is statistically significant.

Example: Suppose we have the datasets:

	1	2	3	4	5	6	7	8	9	10	Average	s(x)
1	5.2	6.0	6.8	6.0	5.9	6.3	7.2	8.0	6.8	7.5	6.57	0.85
2	4.8	7.1	6.3	5.5	5.8	4.9	5.6	7.5	5.1	5.5	5.81	0.90
Diff	0.4	1.1	0.5	0.5	0.1	1.4	1.6	0.5	1.7	2.0	0.76	0.93

Assume that the data have been collected from normally distributed datasets and further that the differences between the observations are normally distributed and that the variances of the sample results are not significantly different (see below). The data meets the criteria for Student's t-test. We do not know, however, which was the experimental model, independent or dependent observations, and use the data for two examples.

One is to assume that rows 1 and 2 are from different experiments and the task is to investigate if the averages are different, i.e., 6.57 ± 0.85 from 5.81 ± 0.90. The second is to assume that the first row pertains to results obtained before a treatment and the second to results after the treatment.

In the first case we apply the t_{ind} and in the second the t_{dep}:

$$1. \quad |t_{ind}| = \frac{\bar{x}_1 - \bar{x}_2}{\sqrt{\dfrac{s_1^2}{n_1} + \dfrac{s_1^2}{n_1}}} = \frac{6.57 - 5.81}{\sqrt{\dfrac{0.85^2 + 0.90^2}{10}}} = \frac{0.76}{\sqrt{0.15}} = \frac{0.76}{0.39} = 1.94;$$

$$df = 10 + 10 - 2 = 18$$

From a t-table (e.g., Table 8) we find that the critical t-value for $df = 18$ is 2.10. Since the calculated t-value is 1.94 we conclude that there is no significant difference between the average values in a two-sided test.

TABLE 8 Extract of a t-Table

	Probability of Two-Tail Test			
df	0.1	0.05	0.01	0.001
9	1.83	2.26	3.25	4.78
18	1.17	2.10	2.88	3.92
20	1.17	2.09	2.85	3.85

Secondly we assume that the results are paired:

2. $\left|t_{\text{dep}}\right| = \dfrac{\overline{d}}{\left(s_d/\sqrt{n}\right)} = \dfrac{0.76 \times \sqrt{10}}{0.93} = 2.59; \quad df = 10 - 1 = 9$

The critical t-value for $df = 9$ is 2.26 in a two-sided test with a probability of 5 % (Table 8). Therefore the null hypothesis is violated and we conclude that there is a difference between the series which have something in common.

The example highlights the importance of understanding the experimental model for a correct statistical inference.

The table value can also be retrieved by the EXCEL functions *T.INV* *(probability,df)* or *T.INV.2T(probability,df)* for one- or two-sided solutions, respectively.

6.7 Effect Size

The concept of effect size is meant to illustrate the difference between two averages.

$$Effect\ size = \frac{Average\ of\ test\ data - average\ of\ control\ data}{Standard\ deviation} \qquad (101)$$

For the concept and calculation of effect size as above to be meaningful the data shall be normally distributed. The standard deviation is commonly that of the control data. The effect size is equal to the z-score and can therefore also be expressed in relation to a normal or t-distribution, e.g., an effect size (z-score) of 1.64 would indicate that 95 % of the control group would be below the value of the average of the test group. The effect size is one-sided.

6.8 Categorical Data, χ^2-Test

The Chi-square test, χ^2–test, is used to prove "goodness of fit," i.e., if values agree or if the observed variation is due to chance alone, i.e., if there is no association (dependence) between values and if the response is what was expected. It can be used to compare the frequency in series with what was expected, e.g., throwing a dice. It is used to evaluate survey categorical data, e.g., opinions on various questions in two or several groups. Only frequency data, e.g., numbers (counts), can be used in the test, not percentages or continuous numbers.

The null hypothesis assumes that there is no difference between the results, i.e., no difference compared to the expected results. The Chi-square test thus tests the probability of independence of a distribution of data and thus excludes that the variation is due to chance alone.

Chi-square is the sum of the differences between found and expected number of observations squared, divided by the expected number:

$$\chi^2 = \frac{(O_1 - E_1)^2}{E_1} + \cdots + \frac{(O_n - E_n)^2}{E_n} = \sum_{i=1}^{n} \frac{(O_i - E_i)^2}{E_i} \quad (102)$$

where n is the number of "groups," O_i the observed sample counts of individuals, and E_i the expected counts.

The expected values can theoretically be any values, as long as their sum equals that of the experiment (study), but they are often estimated in relation to the sample size and distribution between categories and groups.

Example: A classic example is the evaluation of a dice—fair or false? Suppose, after 60 tries, you have obtained the following numbers (frequency) in each "group" (n) 1(8), 2(5), 3(12), 4(9), 5(17), 6(9). Is the dice fair?

The expected number in each group is $60/6 = 10$.

$$\chi^2 = \sum_{i=1}^{n} \frac{(O_i - E_i)^2}{E_i} = \frac{(8-10)^2}{10} + \frac{(5-10)^2}{10} + \cdots + \frac{(17-10)^2}{10} + \frac{(9-10)^2}{10} = 8.4$$

$df = 6 - 1 = 5$. Critical $\chi^2 = 11.1$. Therefore the null hypothesis is accepted, i.e., the differences depend on chance and the dice is fair.

Example: In a survey men and women aged 21−40 and 41−60 were given four choices A, B, C, and D. The outcome was presented in a table:

	A	B	C	D	Total
Women ≤ 40	40	32	16	12	100
Women > 40	28	25	32	15	100
Men ≤ 40	23	12	15	10	60
Men > 40	15	18	10	17	60
Total	106	87	73	54	320

If the expected counts should reflect the distribution of the totals in the groups and categories then the number is calculated from the number in each group, the number in each category, and the total number of observations:

$$E_{c,r} = \frac{C_{c,tot} \times C_{r,tot}}{C_{tot}}, \quad \text{e.g.,} \quad \text{for} \quad \text{men} > 40 \quad \text{in} \quad \text{category} \quad \text{A:}$$

$$E_{A,4} = E_{A,3} = \frac{106 \times 60}{320} = 19.875 \text{ (correctly rounded to 20) and in category}$$

B: $E_{B,4} = E_{B,3} = \frac{87 \times 60}{320} = 16.313$ (rounded to 16).

Theoretically the expected number could be chosen arbitrarily as long as their total sum is the same as in the study.

The degrees of freedom is $(c - 1) \times (r - 1) = 9$.

TABLE 9 The Chi-Square is used in evaluating a 2 × 2 contingency table

	Data Type 1	Data Type 2	Total
Category 1	a	b	a + b
Category 2	c	d	c + d
Total	a + c	b + d	a + b + c + d = N

Inserting these values in the formula gives $\chi^2 = 19.88$. The critical $\chi^2 = 16.92$ ($df = 9$), $p = 0.05$. Thus, the null hypothesis is rejected and the observed difference cannot be caused by chance alone, with a probability of 95 %.

The critical χ^2–value is calculated in EXCEL-$CHIINV(\alpha, df)$ (Table 9). In a 2 × 2 contingency table the df is $(2 - 1) \times (2 - 1) = 1$.

Example A: The true positives (TP), false negatives (FN), true negatives (TN), and false negatives (FN) of a test were calculated and reported in a 2 × 2 table (expected in brackets).

	Quant A		
	Neg	Pos	Tot
Diseased	22 (55.8)	58 (24.2)	80
Nondiseased	257 (223.2)	63 (96.8)	320
Sum	279	121	400

Calculate the χ^2–value.

The expected frequencies will be $\dfrac{121 \times 80}{400} = 24.2$ (diseased, pos) and $\dfrac{279 \times 320}{400} = 223.2$ (nondiseased, neg),

$$\chi^2 = \frac{(58 - 24.2)^2}{24.2} + \frac{(22 - 55.8)^2}{55.8} + \frac{(63 - 96.8)^2}{96.8} + \frac{(257 - 223.2)^2}{223.2} = 84.6;$$

$df = 1$; $p = 0.05$; $\chi^2 - \text{crit} = 3.84$

The critical χ^2-value is 3.84 ($p = 0.05$) and therefore the null hypothesis is rejected and it is not likely that the differences are due to chance but are significantly different.

Example B: A classic example of the use of 2 × 2 tables is the evaluation of treating patients. Assume a study of treating hyperlipidemia with drug A and drug B. To explore any differences using a balanced study, i.e., the same number of participants in each group, which in this case are the independent variables; we can use one of the results as the observed and the other as the expected frequency in the cells of the table.

The expected number of individuals treated by drugs A(x) and B(y)

$$\frac{x}{150} = \frac{240}{500}; \ x = \frac{150 \times 240}{500}; \ x = 72; \quad \frac{y}{150} = \frac{260}{500}; \ y = \frac{150 \times 260}{500}; \ y = 78$$

In the "observed" group the drugs A(z) and B(v):

$$\frac{z}{350} = \frac{240}{500}; \ z = \frac{350 \times 240}{500}; \ z = 168 \quad \text{and} \quad \frac{v}{350} = \frac{260}{500}; \ v = 182$$

	Hyperlipidemia (Expected)			Hyperlipidemia (Observed)		
	"pos"	"neg"	Tot	"pos"	"neg"	Tot
Drug A	72	168	240	130	110	240
Drug B	78	182	260	220	40	260
Sum	150	350	500	150	350	500

The rows and columns add up to the same number and the χ^2 calculated

$$\chi^2 = \frac{(110-72)^2}{72} + \frac{(40-78)^2}{78} + \frac{(130-168)^2}{168} + \frac{(220-182)^2}{182} = 55.1; \quad df = 1$$

Since the χ^2 value is far above the critical value (3.84) the null hypothesis is discarded and there is a difference between the effects of the drugs.

There is a shortcut method for estimating the χ^2-value from a 2 × 2 table:

$$\chi^2 = \frac{(a \times d - b \times c)^2 \times N}{(a+c) \times (a+b) \times (b+d) \times (c+d)} \tag{103}$$

Applying this formula to example A yields a χ^2-value of 84.60.

The χ^2 test for small numbers of observations can be improved by Yates' continuity correction, which gives smaller values of χ^2:

$$\chi^2 = \sum_{i=1}^{n} \frac{(|O_i - E_i| - 0.5)^2}{E_i}; \quad df = n - 1 \tag{104}$$

The difference is usually small enough not to change the conclusion and in practice the Yates' correction has little influence on the outcome unless the total number of observations is less than 40.

The χ^2 evaluation of a 2 × 2 table is an approximation and the exact value can be calculated using Fisher's exact probability test which is includes factorials:

$$p = \frac{\binom{a+b}{a} \times \binom{c+d}{c}}{\binom{N}{a+c}} = \frac{(a+b)! \times (c+d)! \times (a+c)! \times (b+d)}{a! \times b! \times c! \times d! \times N} \tag{105}$$

This calculation becomes exceedingly computer-intensive with large counts.

NB: The previous calculations in these sections have given the $\chi^2-value$ whereas the Fisher's exact test gives the *probability directly*.

6.9 CHI-Square (χ^2), in Comparisons

The imprecision of a measurement method can be compared with the specification for the method:

$$\chi_c^2 = (n-1) \times \frac{s^2}{\sigma^2} \approx (n-1) \times \left(\frac{s}{s_0}\right)^2 \tag{106}$$

where s_0 is the specified (nominal) standard deviation and s the standard deviation found for the measurement procedure.

Example: In a verification procedure the standard deviation of 10 repeated measurements was 0.25 mmol/L. The manufacturer claimed an uncertainty of 0.20 mmol/L. Is the found standard deviation reasonable at a level of confidence of 95 % or $\alpha = 0.05$?

$$\chi_c^2 = (n-1) \times \left(\frac{s(x)}{s_0}\right)^2 = (9) \times \left(\frac{0.25}{0.20}\right)^2 = 14.1 \text{ (rounded 14); The } \chi_{crit(0.05,9)}^2 = 16.9$$

Since the calculated value does not exceed the critical table value the claim is accepted. If, however the standard deviation was based on 30 repeated measurements, the χ_c^2 and $\chi_{crit(0,05,14)}^2$ equal 45.3 and 42.6, respectively and the claim would be rejected. The rationale is that in the latter case the standard deviation would have been estimated with a considerably smaller confidence interval.

In EXCEL, the $\chi_{c(\alpha;(n-1))}^2$ is calculated as *CHIINV(α,df)*.

6.10 Comparison of Variances

One requirement for the proper use of Student's independent t-test is that the variances of the datasets are identical or close to identical. To answer the question if there is a significant difference between the variances of the datasets the F-test is used:

$$F = \frac{s_1^2}{s_2^2} \tag{107}$$

The F-test is also used to answer the general question of whether a dispersion of a normally distributed dataset is significantly different from that of another:

$$F_c = \frac{s_1^2}{s_2^2} = \frac{\text{Var}_1}{\text{Var}_2} \tag{108}$$

The idea of the F-test is that the more the ratio deviates from 1 the larger is the probability is that the variances are different. It facilitates the interpretation of the F-test if the larger variance (Var_1) is in the numerator, thus the F-value is always ≥ 1.

The null hypothesis is that the variances are equal. Then, if a two-sided test is chosen (probability $1 - \alpha/2$, one-sided $1 - \alpha$) and the F-value is larger than the critical value, then the null hypothesis is rejected, i.e., the variances are different.

The critical F-value is found in an F-table or calculated in EXCEL as $F.INV(\alpha,(n_1 - 1),(n_2 - 1))$. If the calculated F-value is less than the critical value, the null hypothesis is accepted. The probability of the variances are considered significantly different if $F_c > F_{crit}$ (table value or $F.INV(\alpha, (n_1 - 1),(n_2 - 1)))$. This conclusion assumes that the df referring to the larger variance is entered as n_1.

The F-table considers the number of observations in both groups, i.e., the df may be different in the samples ($n_1 - 1$) and ($n_2 - 1$), respectively.

For a one-sided test (95 % probability) use $\alpha = 0.05$ ($1 - \alpha$), for a 2-sided test $1 - \alpha/2$.

Since the F-test is based on variances it assumes a Gaussian distribution of the data.

Example: The variances in the example from the t-test were 1.54 and 0.93 with 10 observations in each series. The F-value is 1.66. In EXCEL: $F.INV$ $(0.05,9,9) = 3.18$ for a one-sided test and thus the null hypothesis is accepted, i.e., the variances are not significantly different.

7.   NONPARAMETRIC COMPARISONS

Nonparametric methods are characterized by not relying on distribution-specific parameters, e.g., averages and standard deviations.

In general, conclusions drawn from nonparametric methods are less powerful than parametric with a known distribution and applicable properties. However, as nonparametric methods make fewer assumptions, they are more flexible, more robust, and sometimes applicable to ordinal data.

The nonparametric test for dependent data, corresponding to the Student t_{dep}, is the Wilcoxon signed rank test. To compare independent data, the Mann-Whitney's U-test corresponds to that of Student's t_{ind}. See also Tukey's rapid test.

Nonparametric procedures to compare results are based on ranking the data. Fortunately, the ranking can be achieved in EXCEL without physically sorting the data by using the *RANK* function. The RANK function in previous versions of EXCEL has been developed into *RANK.EQ* and *RANK.AVG* to accommodate the handling of ties of results. The options are that ties are given the same rank number *(RANK:EQ)*or the average of the rank numbers of the ties *(RANK.AVG)*. The function takes the format *RANK.AVG(ABS (A2),$A$2:$D$9)*, where $A$2:$D$9 is the dataset and ABS(A2), ABS(A3) etc. are the absolute values of the difference between the observations in the dataset.

Example: Rank the data in the table according to the two models and observe the difference

Obs. No:	1	2	3	4	5	6	7	8	9	10	11	SUM
VALUES	2	2	3	5	5	5	5	7	8	10	11	
RANK.AVG	1.5	1.5	3	5.5	5.5	5.5	5.5	8	9	10	11	66
RANK.EQ	1	1	3	4	4	4	4	8	9	10	11	59

RANK.AVG will give the average of tied results, however in the calculation the first and last of a group of even number of tied values are disregarded, unless the number of observations is two. *RANK.EQ* gives the same rank number to all observations in a tied group, namely the lowest. **NB:** the rank refers to the observation number; therefore it is important to arrange the observation in ascending or descending order if ranking is done manually. Both ranking orders are correct but may have different uses. The sum of the ranks differs, thus the rank sum of *RANK.AVG* is constant and equal to the sum of the number of observations whereas the rank sum of *RANK.EQ* varies depending on the number of tied values.

7.1 Wilcoxon Signed Rank Test for Paired Samples

Confusingly, there are two Wilcoxon rank tests, the *signed rank test* and the *rank sum test*. A confusing fact is that the latter is sometimes called the rank sign test. The *signed rank test* is used for paired observations similar to the Student's t_{dep}-test whereas the *sum rank test* is used for unpaired or independent results, i.e., evaluating the difference between medians (not averages like the Student's t_{ind}). We will only address the *Wilcoxon signed rank test* here.

The procedure is to rank the differences between the pairs, irrespective of their signs. Any ties are usually given the average of the ranks. Then the positive and negative differences are identified and added to give two numbers T^+ and T^-. These are the positively and negatively signed summed ranks, respectively. The smaller of these indicators is compared with the proper table.

The null hypothesis of the test is that there is no difference between the paired observations or, in other words, that the sum of the ranked pairs in the positive and negative groups are equal.

The smaller of T^+ and T^- is used for the evaluation. A dedicated table should be used. If the number of paired observations (n) is large, e.g., >20, a normalization (Gaussian approximation) can be used

$$\mu_T = \frac{n(n+1)}{4} \tag{109}$$

and

$$\sigma_T = \sqrt{\frac{n \times (n+1) \times (2n+1)}{24}} \tag{110}$$

The difference can then be expressed as a z-score:

$$z = \frac{T_{\min} - \mu_T}{\sqrt{\frac{n \times (n+1) \times (2n+1)}{24} - q}} \tag{111}$$

$$q = \sum \frac{t^3 - t}{48}$$

The z-value can be evaluated using an ordinary t-table but it is often suggested to use $z > 1.96$ irrespective of the number of observations. The significance may be more significant than what the table suggests and therefore a modifying term q is introduced in Eq. (111). Each group of tied ranks is included in the sum, e.g., if two number are tied $t = 2$ if three the $t = 3$.

A rule of thumb is that the larger the difference between the T^+ and T^- the more probable is the significance of the difference.

The maximum sum of the possible ranks is always equal to

$$\frac{n \times (n+1)}{2} \tag{112}$$

In the example above there were 11 observations and accordingly the maximum rank sum was 66.

Example: The following paired data are extracted from a comparison of two measuring procedures. Evaluate the difference between results.

1. Calculate the difference as a decrease between what is appointed as the first measurement and the second but keep track of the sign. Thus, a decrease is a negative number.
2. Ignore the sign and rank the absolute difference. Any differences that are 0 should be disregarded.
3. Ties should be given the same numbers and the remaining updated accordingly.
4. Find the sum of the positive ranks and the negative ranks.

	1	2	3	4	5	6	7	8	9	10	11	12	13	14	15
Assay 1	1.24	1.34	1.39	1.41	1.64	1.44	1.48	1.51	1.54	1.54	1.54	1.62	1.63	1.65	1.70
Assay 2	1.30	1.50	1.70	1.50	1.44	1.47	1.60	1.60	1.80	1.50	1.70	1.90	1.81	1.70	1.65
Diff	−0.06	−0.16	−0.31	−0.09	0.20	−0.03	−0.12	0.09	−0.26	0.04	−0.16	0.28	−0.18	−0.05	0.05
Abs diff	0.06	0.16	0.31	0.09	0.20	0.03	0.12	0.09	0.26	0.04	0.16	0.28	0.18	0.05	0.05
Rank	5	9.5	15	6.5	12	1	8	6.5	13	2	9.5	14	11	3.5	3.5

(T^+) is 38 and (T^-) 82; Thus, $(T^+) + (T^-)$ is $\dfrac{12 \times (12+1)}{2} = 120$.

$$\mu_T = \frac{15 \times (15+1)}{4} = 60; \quad \sigma_T = \sqrt{\frac{15 \times (15+1) \times (2 \times 15 + 1)}{24}} = \sqrt{\frac{60 \times 31}{6}} = 17.61;$$

$$q = \sum \frac{t^3 - t}{48} = \frac{2^3 - 2 + 2^3 - 2 + 2^3 - 2}{48} = 0.375; \quad z = \frac{82 - 60}{\sqrt{17.61^2 - 0.375}}$$

$$= 1.25 \text{ (rounded 1.3)}.$$

Thus the difference between the pairs would be not reach significance with a $p < 0.05$ ($p = 0.016$; 2-tail).

7.2 Mann-Whitney Test for Unpaired Samples

The preferred nonparametric method for unpaired samples is the Mann-Whitney test or Mann-Whitney U-test and thus the nonparametric solution to evaluating two independent datasets comparable to the Student's t_{ind}. Essentially the significance of a difference of the location (value) of the medians of two groups is tested. In the test all the results are ranked as if they belonged to one series of measurements and then the ranks of the samples belonging to each of the two methods are added separately.

There are different means to evaluate the results. The easiest is that advised by Wilcoxon, by which a test statistic T_S is the sum of the ranks in the smaller group, i.e., the group with the fewer observations (if groups are equally sized then either can be chosen). The statistic is evaluated in a special Mann-Whitney table.

Other statistics may be calculated and used for evaluation of the comparison. The sum of the ranks are T_S and T_L representing the smaller and larger rank sums, respectively. If the groups were the same the difference in rank sums should be small.

A test statistic U is calculated as:

$$U = T_s - \frac{n_a \times (n_a + 1)}{2} \tag{113}$$

where n_a refers to the number in the group with the lower rank sum. Consult a "Mann-Whitney table" which usually has the number of observations in both groups as entries. The target is to find the last "probability column" that does not contain the value of the statistic (U). There are different layouts of the tables but they only cover up to a total number of observations of about 20. If there are more observations, the rank sum approaches a normal distribution and a z-value can be calculated. Two approaches are available; although they look different the outcome is usually the same (the literature sources are Altman, Practical statistics in Medical Research and Engineering Handbook (NIST)):

$$\mu_s = \frac{n_s(N+1)}{2} \quad \text{(Altman)} \tag{114}$$

Alternatively:

$$\mu_s = \frac{n_s n_l}{2} \quad \text{(Engineering Handbook)} \tag{115}$$

$$\sigma_s = \sqrt{\frac{n_s \times n_l \times (N+1)}{12}} \tag{116}$$

$$z = \frac{T_s - \mu_s}{\sigma_s} \quad \text{(Altman)} \tag{117}$$

$$z = \frac{U_s - \mu_s}{\sigma_s} \quad \text{(Engineering Handbook)} \tag{118}$$

Where n_s is the number of observations in the group with the lower rank, n_l the number of observations in the other group, T_s is the lower rank sum, and U_s is the statistic calculated according to Eq. (113). The z-value can then be evaluated by a standard normal distribution or *NORM.S.DIST(z,true)*.

Example: The concentration of two different materials was measured by the same procedure.

	1	2	3	4	5	6	7	8	9	10	11	12	
Material 1	1.24	1.34	1.39	1.41	1.64	1.44	1.48	1.51	1.54	1.54	1.54	1.62	
Material 2	1.30	1.50	1.70	1.50	1.44	1.47	1.60	1.60	1.80	1.50	1.70	1.90	Sum
Rank	1	4	5	6	3	7	10	14	15	16	17	19	117
	2	12	21.5	12	8	9	20	18	23	12	21.5	24	183

$$U_s = 12 \times 12 + 0.5 \times 12 \times 13 - 117 = 105; \quad U_s + U_L = \frac{24 \times (25+1)}{2} = 300$$

$$\mu_s = \frac{12 \times (24+1)}{2} = 150 \quad \text{(Altman)}$$

$$\mu_s = \frac{12 \times 12}{2} = 72 \quad \text{(Engineering Handbook)}$$

$$\sigma_s = \sqrt{\frac{12 \times 12 \times (24+1)}{12}} = 17.32$$

$$z = \frac{|117 - 150|}{17.32} = 1.91 \quad \text{(Altman)}$$

The *NORM.S.DIST* gives $p = 0.03$

$$z = \frac{|105 - 72|}{17.32} = 1.90; \quad \text{(Engineering Handbook);} \quad p = 0.03.$$

Note that if the larger or smaller U is evaluated, $1 - \text{NORM.S.}$ DIST $= 0.03$.

The probability can also be estimated from the test statistic (U). A selection of lines from an appropriate Mann-Whitney table indicates the probability:

n_1	n_2	$p = 0.1$	$p = 0.05$	$p = 0.02$	$p = 0.01$
11	12	104−160	99−165	94−170	90−174
12	12	120−180	115−185	**109−191**	105−195
12	13	125−187	119−193	113−199	109−203

In this example there are 12 observations in each group and the test statistic U_s is 105. The last column of the table that *does not* include the test statistic is $p = 0.02$.

An alternative to identify a difference between independent samples is the *Tukey's quick test* or *"Total end count."* This test is distribution free and tests if ranked datasets are different by counting the number of observations in dataset B below those in dataset A and the number of results in dataset A above those in dataset B. Thus, the total number of observations in the series which are not in the overlap of the series are counted. The sum of the number of observations identified this way is the "end count," which is the metric; if the end count is 7 the datasets are significantly different on the 95 % level, if 10, 99 %. The limitation is that the number of observations in the groups shall be "about" the same but the total number of observations is not important and the overall maximum and minimum numbers must not be in the same dataset.

8. ANALYSIS OF VARIANCE

8.1 Definitions and Calculation

The one-way analysis of variance (ANOVA) is designed for comparing averages of several of datasets. It should be used rather than using a series of Student's independent t-tests for all possible pairs which would overestimate the differences due to chance. The ANOVA procedure will evaluate if there is a difference between the averages of the studied groups but not indicate which observation(s) is deviating. It is still possible to infer differences between individual groups.

TABLE 10 Standard ANOVA output table

Between groups	SS_b	df_b	MS_b	$F = MS_b/MS_w$	p-value
Within groups	SS_w	df_w	MS_w		
Total	SS_{tot}	df_t			

TABLE 11 Notations used in describing an ANOVA

	Group 1	Group 2	...	Group k	
Result 1:	x_{11}	x_{21}	...	x_{k1}	
Result 2:	x_{12}	x_{22}	...	x_{k2}	
⋮	⋮	⋮	⋮	⋮	
Result n:	x_{1n}	x_{2n}	...	x_{kn}	
Average:	$\bar{x}_1$	$\bar{x}_2$		$\bar{x}_k$	$\bar{\bar{x}}_{n \times k}$
Std dev:	s_1	s_2		s_k	$s_{n \times k}$

The ANOVA calculates the sum of squares within the groups and between the groups. The significance of a difference is established applying F-statistics (Eq. 108) to the ratio between variances (mean squares). The ANOVA procedure is offered by all standard statistical packages and built into many spreadsheet programs. The solution to an ANOVA is often presented in a standardized format (Table 10):

where SS is the "sum of squares," df degrees of freedom, and MS "mean square." The F-value indicates the significance of the difference between groups in the study (Table 11).

The experimental design comprises several (k) groups, runs or series of results (x_i), each including several (n_i) observations or results with a total number of $N = k \times n$ observations. The calculation of the standard deviation of the entire dataset is described in the section "Analysis of variance components."

The null hypothesis is "there is no difference between the averages of the groups."

Groups may comprise different numbers of observations or results (unbalanced design). The group averages are designated $\bar{x}_k$ and the grand average $\bar{\bar{x}}$.

There are different ways to calculate the parameters of an ANOVA but all are based on calculating the sum of squares (SS) for the total, within and between groups.

Total sum of squares (SS_{tot}):

$$SS_{tot} = \sum_{i=1}^{i=N} (x_i - \bar{\bar{x}})^2 = (N-1) \times Var(x_i) = \sum_{i=1}^{i=N} x_i^2 - \frac{\left(\sum_{i=1}^{i=N} x_i\right)^2}{N}$$
$$= \sum_{i=1}^{i=N} x_i^2 - N \times \bar{\bar{x}}^2;$$

(119)

$$df_{tot} = N - 1 \tag{120}$$

Between-groups sum of squares:

$$SS_b = n_0 \times \sum_{i=1}^{i=k} \left(\bar{x}_i - \bar{\bar{x}}\right)^2 \tag{121}$$

where n_0 is the number of observations in the groups. If this varies between groups (an unbalanced design) then Eq. (121) is rearranged to

$$SS_b = n_0 \times \sum_{i=1}^{i=k} \left(\bar{x}_i - \bar{\bar{x}}\right)^2 = \sum_{i=1}^{i=k} \left(n_i \times (\bar{x}_i - \bar{\bar{x}})^2\right)$$
$$= \sum_{i=1}^{i=k} n_i \times \bar{x}_i^2 - \frac{\left(\sum_{i=1}^{i=N} x_i\right)^2}{N};$$

(122)

$$df_b = k - 1 \tag{123}$$

If the design is unbalanced, use a "weighted" grand average

$$\bar{\bar{x}} = \frac{\sum_{i=1}^{i=k} n_i \times \bar{x}_i}{\sum_{i=1}^{i=k} n_i} = \frac{\sum_{i=1}^{i=k} n_i \times \bar{x}_i}{N} \tag{124}$$

Within-groups sum of squares:

$$SS_w = (N-k) \times \frac{(n_1 - 1) \times s_1^2 + \cdots + (n_k - 1) \times s_k^2}{n_1 + \cdots + n_k - k}; \tag{125}$$

$$df_w = N - k \tag{126}$$

Compare the calculation of SS_w with the that of the pooled standard deviation (Eq. 30)!

If the groups comprise the same number of observations (balanced) then

$$SS_w = \sum_{i=1}^{i=k} \left((n_i - 1) \times s_i^2\right) = \sum_{i=1}^{n_i} (x_i^2) - \sum_{i=1}^{n_i} n_i \times \bar{x}_i^2$$
$$SS_{tot} = SS_b + SS_w \tag{127}$$

It may be practical to use the simplest of the formulas for SS_{tot} among those in Eq. (119) and subtract either SS_w or SS_b to calculate the SS_b and SS_w, respectively.

8.2 ANOVA to Evaluate Differences Between Averages

To evaluate the difference between the averages of several groups uses the F-test. In the ANOVA evaluation the F is calculated as

$$F = \frac{\text{Var}_b}{\text{Var}_w} = \frac{\text{MS}_b}{\text{MS}_w} \tag{128}$$

The MS_b shall be in the numerator and MS_w in the denominator. This is the order irrespective of the relative sizes of the MS. The F is evaluated using a normal F-table; in EXCEL use $F.INV(prob, df_1, df_2)$.

A high F-value indicates a significant difference of the averages of groups but not between *which* groups. There are different techniques to estimate the significance level of differences between groups.

A simple approach is to arrange the averages in increasing or decreasing order and calculate the "least significant difference":

$$\Delta_{\text{sign}} = t_{(n-2)} \times s_{\text{within}} \times \sqrt{\frac{2}{n}} \tag{129}$$

where s_{within} is the estimated within group standard deviation $\left(\sqrt{\text{MS}_{\text{within}}}\right)$, n is the number of observations in each group (balanced design), $t_{(n-2)}$ the t-value for the indicated degrees of freedom.

Comparison of this value with the difference between the ordered averages will indicate where a significant difference may be found. This formula is derived from Eq. (96) (independent Student's t), assuming that the standard deviation is the same for the groups and the difference between the averages is the difference that is tested for significance. This is directly seen by rearranging the formula to

$$t_{(n-2)} = \frac{\Delta_{\text{sign}}}{\sqrt{\frac{2 \times s_{\text{within}}^2}{n}}} \tag{130}$$

A formula applicable in unbalanced designs with different standard deviations would then be

$$\Delta = t_{(n-2)} \times \sqrt{\frac{s_1^2}{n_1} + \frac{s_2^2}{n_2}} \tag{131}$$

Remember that the standard deviation of the groups needs to be similar and the degrees of freedom may need to be calculated using the Satterthwaite's approach (see Eq. 99).

There are more sophisticated and rigorous solutions to this problem.

Example: The S-Cholesterol concentration was measured in 10 participants on four different diets:

	1	2	3	4	5	6	7	8	9	10	Average	Var
1	5.2	5.6	6.8	3.5	5.9	6.3	7.2	8.5	6.8	5.7	6.15	1.77
2	6.2	6.2	7.9	8.2	5.7	6.6	8.1	9	12.1	6.7	7.67	3.57
3	4.2	4.9	6.8	3.7	4.5	5.6	6.4	5.8	6	7	5.49	1.26
4	4.8	7.1	5.9	4.7	5.8	4.9	5.6	7.5	5.1	5.8	5.72	0.89

"Grand average": 6.26, Tot variance: 2.47:

We can now determine the variance between individuals (columns) and between diets (rows). The "groups" can be either participants or diet, compared to our definitions above it is equal to turn the table 90 degrees. It is important for the evaluation to remember that the null hypothesis states that there is no difference between the groups.

Diet variation, rows (null hypothesis: there is no difference between averages of the concentration of S-Cholesterol in the groups, i.e., by diet):

SS_{tot}: $(40 - 1) \times 2.47 = 96.50$; $df = 40 - 1 = 39$.

SS_b: $10 \times [(6.15 - 6.26)^2 + (7.67 - 6.26)^2 + (5.49 - 6.26)^2 + (5.72 - 6.26)^2] = 28.85$; $df = 4 - 1 = 3$.

Alternatively $96.50 - 67.65 = 28.85$

$SS_w = (10 - 1) \times (1.77 + 3.57 + 1.26 + 0.89) = 67.65$; $df = 40 - 1 - 3 = 36$.

Alternatively $96.50 - 28.85 = 67.65$

$$F = \frac{SS_b}{SS_w} \times \frac{df_w}{df_b} = 5.12.$$

The critical F-value is found in a table or by EXCEL: $F.INV([1 - \alpha], df_b, df_w, true)$ and the corresponding p-value.

The F-test is one-sided and the critical value is the area under the F distribution curve which is to the right of your stated probability. Likewise when the p-value is calculated, it is counted from the right end of the distribution, hence $p = 1 - F.DIST(F\text{-}value; df_b; df_w; treu)$. The larger the F-value the less probable it is that the group averages are equal, and the null hypothesis is thus rejected.

The ANOVA table from in EXCEL:

	SS	df	MS	F	p-Value	F-crit
Between groups	28.85	3	9.62	5.12	0.005	2.87
Within groups	67.65	36	1.88			
Total	96.50	39				

The result indicates that there is a significant difference between the diets.

It is convenient but not necessary to sort the calculated averages in ascending order: 5.49, 5.72, 6.15, 7.67.

Since the variances (standard deviations) of the groups are different, formula (131) is applicable.

$$\Delta_{sign} = t_{0.05,18} \times \sqrt{\frac{3.57}{10} + \frac{0.89}{10}} = 2.10 \times 0.668 = 1.40$$

A "cross table" displays and summarizes the differences between the averages

	5.49	5.72	6.15	7.67
5.49	0	0.23	0.66	2.18
5.72		0	0.43	1.95
6.15			0	1.52
7.67				0

Thus, there are significant differences (>1.4) between the highest and the remaining three groups between which there is no significant difference in S-Cholesterol concentrations related to the diet.

Variation between the individuals:

The corresponding estimate of variation between diets:

	SS	df	MS	F	p-Value	F-crit
Between groups	32.13	9	3.57	1.66	0.142	2.21
Within groups	64.37	30	2.15			
Total	96.50	39				

SS_{tot}: 96.50; $df = 40 - 1 = 39$.

SS_b: 32.13; $df = 10 - 1 = 9$.

Alternatively $96.50 - 64.37 = 32.13$

$SS_w = 64.37$; $df = 40 - 1 - 9 = 30$.

Alternatively $96.50 - 32.13 = 64.37$

$$F = \frac{SS_b}{SS_w} \times \frac{df_w}{df_b} = 1.66.$$

Since the calculated F-value is less than the critical value (2.21), there is no difference between the S-Cholesterol concentrations among the individuals.

8.3   Nonparametric Methods

Use of ANOVA requires that the data are normally distributed and that the variances within the groups are of a similar magnitude (cf. Student's t-test) in the measuring interval.

The *Kruskal-Wallis test* is a nonparametric alternative to the one-way ANOVA and the *Friedman's test* can be compared with the two-way ANOVA. In both tests all the observations are ranked together, usually any ties given the same calculated rank, i.e., *RANK.EQ* in EXCEL. Then the sum of the ranks in each method are separated and used to calculate statistics that can be evaluated by comparing with a χ^2 table.

Both procedures can be applied to ordinal, interval, and rational data. The test will only demonstrate that there is at least one group which differs from the rest. Post-test tests may be required to identify significant differences.

9. ANALYSIS OF VARIANCE COMPONENTS

A common problem in applied analytical chemistry is to identify sources of uncertainty in measurement procedures, in particular to identify the variance that can be attributed to variation between runs and within runs, however a run, or series, is defined. The experimental setup to solve this is to measure the same sample repeatedly in several series. This creates sufficient information to use in an ANOVA and evaluate the mean squares to estimate the within- and between series variances and further the combined or intralaboratory variance.

The mean squares (MS) of the ANOVA table are obtained by dividing the sum of squares by the corresponding degrees of freedom:

$$MS_w = \frac{SS_w}{df_w} \tag{132}$$

$$MS_b = \frac{SS_b}{df_b} \tag{133}$$

The *within series variance* (MS_w) is equivalent to the pooled average of the variances of the results of the groups. Therefore, if the study is balanced the MS_w is the average of the variances of the individual series.

The *between series variance* can be calculated as the average number of observations (n_0) in each group times the variance of the group averages, s_g^2, i.e., involving the "gross average":

$$s_g^2 = \frac{\sum_{i=1}^{i=k} \left(\bar{x}_i - \bar{\bar{x}}\right)^2}{(k-1)} \tag{134}$$

$$MS_b = n_0 \times \frac{\sum_{i=1}^{i=k} \left(\bar{x}_i - \bar{\bar{x}}\right)^2}{(k-1)} = n_0 \times s_g^2 \tag{135}$$

If, however, the experiment is not balanced the number of observations in each group needs to be included.

$$MS_b = \sum_{i=1}^{i=k} \left(n_i \times \bar{x}_i^2\right) - N \times \bar{\bar{x}}^2$$

The total (combined) variance can only be estimated after compensation for the contribution from the within series variance to the MS_b:

"*Purified*" (or "*pure*") between run or intermediary precision:

$$s_b = \sqrt{\frac{MS_b - MS_w}{n_0}} = \sqrt{s_g^2 - \frac{s_{pool}^2}{n_0}} \tag{136}$$

where n_0 is the average number of observations in each group, run, or series.

s_b^2 is also known as the "unbiased estimate of the between group variance."

If we have an unbalanced setup, then a "harmonic" average number of observations should be used:

$$n_0 = \frac{N^2 - \sum_{i=1}^{i=k}(n_i)^2}{N \times (k-1)} = \frac{N - \frac{\sum_{i=1}^{i=k} n_i^2}{N}}{k-1} = \bar{n}_i - \frac{s(n)^2}{N} \qquad (137)$$

where N is the total number of observations and n_i is the number of observations in each group and k is the number of groups. In most cases the difference between the arithmetic mean of the number of observations and the "harmonic" average is negligible, i.e., $Var(n)/N$, *the second term in formula 139*, is comparatively small in practical work.

9.1 Combined Uncertainty

$$s_{tot} = u_c(x) = \sqrt{(s_b)^2 + MS_w} \qquad (138)$$

In most cases MS_b would not be assumed to be less than the MS_w, since the within group variance is a "constituent" of the between group variance. This would also violate the formula (136) which suggests the square root of a negative number. In practice, however, it happens that $MS_b < MS_w$. In those cases, by convention, the s_{tot}, is set to $\sqrt{MS_w}$, i.e., $MS_b = 0$. This condition can be formulated as

$$s_b^2 = MAX\left(0, \frac{MS_b - MS_w}{n_0}\right) \qquad (139)$$

In EXCEL this function will return the maximum value of those within the brackets, i.e., it is either a positive value or zero.

The total variance of several series of values can be calculated by different approaches, for instance, the total estimated variance of results from a laboratory with several instruments performing the same measurements; i.e., the s_{tot} could be the estimated square root of the variance from the total dataset $\sqrt{var(x_{11}:x_{kn})}$, k representing the number of series and n the number of observations in the series, thus $x_{11}:x_{kn}$ represents all observations from the first to the last.

It can be argued that a representative variance is the average of the variances of the series.

The s_{tot} can also be estimated by the analysis of variance components. The difference between these approaches is small if there is no or a very small (less than about 1 %) "between series" variation.

Example: Control material was measured five times in five series (Table 12). Evaluate any difference between the averages of the series and calculate the within-, between-, and combined uncertainties!

$$s_b^2 = \frac{(23.7 - 6.88)}{5} = 3.36; \quad s_b = 1.83$$

$$s_w^2 = 6.88; \quad s_w = 2.62$$
$$u_c(x) = \sqrt{6.88 + 3.36} = \sqrt{10.24} = 3.20$$

$$\sqrt{VAR(x_{11}:x_{kn})} = \sqrt{9.66}; \quad s(x) = STDEV(x_{11}:x_{kn}) = 3.11$$

Conclusion: There is a significant difference between the series ($F = 3.4$). The major source of the combined uncertainty is the within series variation. The total variance differs considerably between the two different methods for its estimation (Table 13).

A combined (instrument, total, within laboratory) uncertainty ($s(x_i)$) is necessary to compare the performance of for instance two measurement

TABLE 12 Results of repeated measurements

	Groups 1	Groups 2	Groups 3	Groups 4	Groups 5
Result 1	123	118	123	118	125
Result 2	125	121	120	125	128
Result 3	121	127	123	125	129
Result 4	126	125	125	118	127
Result 5	124	122	123	120	126
Average	128.8	122.6	122.8	121.2	127.0
Variance	3.7	12.3	3.2	12.7	2.5

TABLE 13 Standard ANOVA table

Source of Variation	SS	df	MS	F	p-Value
Between groups	94.64	4	23.66	3.4	0.03
Within groups	137.6	20	6.88		
Total	232.2	24			

procedures. It shall be expressed as the standard error ($s(\bar{x})$), i.e., $s(\bar{x}) = s(x_i)/df_{s(\bar{x})}$.

10. REGRESSION

Regression is the statistician's term for describing the relation (dependence, association) between two variables. By convention the independent (reference or comparative) variable is shown on a horizontal axis (the X-axis) and the dependent (test) variable on the vertical (Y-axis). It may be useful to visualize the independent variable as the cause and the dependent as the effect variable. This is a common terminology, particularly in multivariate analysis, but it should be emphasized that this does not imply that the regression will address the causality of a found association.

Regression is also the prediction of a value from another variable.

In analytical work regression analysis is used for calibration functions. In calibrations the value of the calibrator is the independent variable and the signal the dependent variable. Regression is also used in comparison of results of two measurement procedures.

To establish a regression function the two quantities are measured in the same sample and thus pairs of values are obtained which can be represented in a two-dimensional diagram, often recognized as a "scattergram" (Fig. 8). The simplest regression function describes a linear (first order, ordinary linear regression, OLR) relationship but there are an unlimited number of types of functions.

The mathematical function that describes a linear or first order relationship (regression) can be established from a minimum of two pairs of observations or from one point, provided the slope is known. Axiomatically one and only one straight line can be drawn between two points.

In comparing results of measurements it is recommended and important to display observations in the scattergram to provide a visual impression of the dataset. This will facilitate recognizing trends, "outliers" and distribution of data points.

10.1 Ordinary Linear Regression

The ordinary linear regression (OLR) or linear least square regression function describes a straight line in a two-dimensional diagram. Its mathematical representation is

$$Y = b \times X + a \qquad (140)$$

where b is the slope of the line and a is the "Y-intercept," i.e., the Y-value where the line crosses the Y-axis, i.e., the value of Y if $X = 0$.

The ORL establishes a line that minimizes the vertical differences between each observation, at each X-value, and the line, thus disregarding

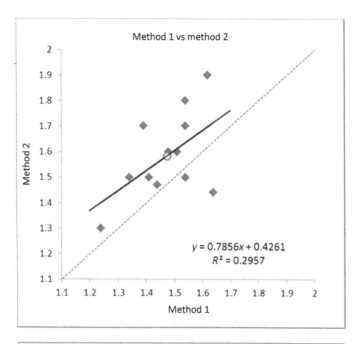

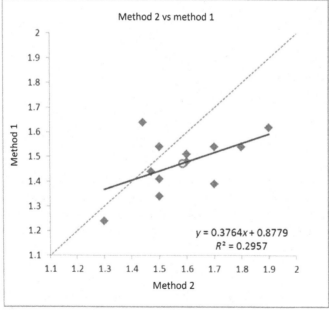

FIGURE 8 Scattergram of results of the example. In the right panel the variables have been swapped, i.e., the Method 1 of the table is on the vertical axis and the Method 2 on the horizontal axis. The function of the ORL is shown and the coefficient of detection ($r^2 = R^2$). **NB:** The functions are different and describe the inverse relation of the variables. The observations are "mirrored" across the equal line. The coefficient of determination is unchanged. The equally sized units and length of axes facilitate comparing the regressions with the "equal line" with a slope of 1, i.e., 45 degrees and intercept 0. The regression lines pass through the average of the values of the dependent and independent variables (shown as a circle in the diagrams).

and uncertainty of the X-value (see Fig. 8). Therefore, it is important to choose the variable with the smaller measurement uncertainty as the independent quantity on the horizontal axis (X-value). A regression line represents a kind of average of all observations in the measuring interval. It is always centered on the average of the independent and average of dependent variables $(\bar{x}/\bar{y})$.

If a set of paired observations (x_i/y_i) and the number of pairs (n) are given, the regression function can be calculated.

First the slope ($b_{y/x}$) or "regression coefficient" is calculated:

$$b_{y/x} = \frac{\sum_{i=1}^{n}[(x_i - \bar{x}) \times (y_i - \bar{y})]}{\sum_{i=1}^{n}(x_i - \bar{x})^2} = \frac{\sum_{i=1}^{n}(x_i \times y_i) - n \times \bar{x} \times \bar{y}}{\sum_{i=1}^{n}x_i^2 - n \times (\bar{x})^2} \quad (141)$$

The formulas can be simplified by defining the "sum of squares":

$$SS_{xx} = \sum_{i=1}^{n}x_i^2 - \frac{\left(\sum_{i=1}^{n}x_i\right)^2}{n} = \sum_{i=1}^{n}(x_i - \bar{x})^2 = (n_x - 1) \times s(x)^2 \quad (142)$$

$$SS_{yy} = \sum_{i=1}^{n}y_i^2 - \frac{\left(\sum_{i=1}^{n}y_i\right)^2}{n} = \sum_{i=1}^{n}(y_i - \bar{y})^2 = (n_y - 1) \times s(y)^2 \quad (143)$$

$$SS_{xy} = \sum_{i=1}^{n}(x_i \times y_i) - \frac{\sum_{i=1}^{n}x_i \times \sum_{i=1}^{n}y_i}{n} = \sum_{i=1}^{n}[(x_i - \bar{x}) \times (y_i - \bar{y})]$$
$$(144)$$

Then,

$$b_{y,x} = \frac{SS_{xy}}{SS_{xx}} \quad (145)$$

This gives the slope of the regression line. Since the line is bound to pass through the average of the variables, the average of the dependent ($\bar{y}$) and independent (reference) ($\bar{x}$), measurements can be used, together with the estimated slope, to estimate the intercept and thus the regression function can be calculated:

$$\bar{y} = b_{y/x} \times \bar{x} + a; \quad a = \bar{y} - b_{y/x} \times \bar{x} \quad (146)$$

$b_{y/x}$ is the slope of the regression function "Y on X." The slope of "X on Y" is written $b_{x/y}$. If the slope is not identified by an index it is usually $b_{y/x}$.

$$b_{x,y} = \frac{SS_{xy} \times SS_{xx}}{SS_{yy}}$$

Since the averages of the quantity values are used in the calculation of the slope and also the intercept, it is important that the averages are representative of the quantity values. This, strictly, requires that the quantity values are normally distributed along both axes.

The model requires that for each value x_i of the independent variable there is a random, i.e., normal, distribution of y values with an average on the estimated regression line. The y_i has been drawn from this distribution and there may therefore be an unlimited number of y-values to each x_i. This further emphasizes the requirement of normally distributed quantity values.

The pair of averages is also recognized as the centroid of the function.

In the OLR the sum of the squared vertical distances between the observations (y_i) and the corresponding points on the regression line ($\hat{y}$), calculated by the regression function are minimized. These distances are recognized as "residuals."

A consequence of the OLR model is that it does not include, or consider, any measurement uncertainty of the independent variable (X).

In trigonometrical terms the slope is the tangent of the angle between the line and the X-axis. The tangent can take any value between 0 and $\pm\infty$ where 0 represents a horizontal line and $\pm\infty$ a vertical. A slope of 1 is 45 degrees, i.e., dividing the quadrant in two equal parts. Any negative slope represents a line that goes from high values of the dependent variable to low values as the independent variable increases.

There are two extremes of linear regression, one is if the line is vertical and the slope is then undetermined; any Y-value corresponds to the only given X-value. The other extreme occurs when there is no variation of the dependent variable as the independent varies. That is to say that the slope is zero or not significantly different from zero. In either case there is no useful relation between the variables.

Even if an estimated slope looks quite different from zero the distribution of the results, e.g., shown by a small correlation coefficient may have a crucial impact on its significance from zero, which becomes important to infer.

For this calculation the standard deviation of the slope is needed. This is readily available in EXCEL if the LINEST command is used. This is a powerful "array" command which reports a table which contains ten items as described in the table below. a worked example is given in the section of linear calibration.

Slope	Intercept
Standard error of slope se(slope)	Standard error of intercept se(intercept)
Coefficient of determination (r^2)	$S_{y,x}$
F-value	df (N−2)
SS$_{regression}$, explained variation	SS$_{residual}$, unexplained variation

One method to evaluate the significance is to calculate a t-value as the slope/[se(slope)] to be evaluated by the T.DIST function (see below, under Calibration). Alternatively, use the F-value of the table and the proper degrees of freedom (1 and $N-2$). The df can be understood from an

ANOVA table. The SS_{tot} equals SS_{yy}, and SS_{regr} is $\left(s_{xy}\right)^2/s_{xx}$ with a df of 1 (for abbreviations see Eqs. 142–144). The SS_{resid} is obtained by subtraction and its $df\, N - (1 + 1)$. The total variation has $df = N - 1$.

	SS	df	MS	F	p-Value
Regression (explained)	SS_{regr}	1	SS_{regr}	$SS_{regr}/$ SS_{resid}	F.DIST(F;df₁, df₂)
Residual (unexplained)	SS_{resid}	$N - 1 - 1$	$SS_{resid}/$ $(N - 2)$		
Total	SS_{tot}	$N - 1$			

The ratio between the SS_{regr} and SS_{tot} is the coefficient of determination (r^2), therefore the r^2 describes the part of the total variation that is explained by the model.

10.1.1 Comments

A characteristic of the OLR, or "method of linear least squares" is that the variance (uncertainty) of measurements of the independent quantity (X-value) is assumed to be zero or that the ratio between $s(y)^2$ and $s(x)^2$, referred to as λ_i, discussed below Eq. (152), is large. The OLR is relatively robust and acceptable results may be obtained also if the variance of X-values is >0. Further, the variance of the independent variable (Y) should be homoscedastic, i.e., the same within the measuring interval.

The OLR is sensitive to outliers and extreme values will therefore have a major impact on the OLR function (see Section 11.10).

The quantity values should ideally be normally distributed in both dimensions and the OLR is regarded as rather robust also in that respect.

The concept "inverse regression" is found in multivariate regression analysis and it appears particularly in discussions on dimension reduction. It is also used for the regression of the format

$$Y = \frac{1}{X} \times b + a$$

which is no longer a linear regression of Y on X. If the regression of Y on X is linear with a positive slope ($b > 0$) the inverse regression is nonlinear and the trend negative.

10.2 Linearity

The linear regression analysis assumes a linear relation, i.e., characterized by a first order function that describes the relation between a dependent and independent variable. In an experimental setting or when a linearity is assumed, or known, the goal of the regression analysis is to establish this relation, e.g., the relation between light absorbance and the concentration of an analyte.

In formal method validation the linearity shall be proven. A thorough procedure for evaluating the linearity of a measurement procedure has been published as EP6 by the CLSI (www.CLSI.org).

If the intercept is zero the function will provide results that are directly proportional to the concentration of the analyte in the sample. A sample that is diluted $1 + 1$ shall therefore give a signal that is half of the original signal and the result shall be half of that the original sample. Serial dilutions can therefore be used to verify the linearity. This is not always the case in biological samples; a reason may be that possible inhibitors become inefficient or more potent in diluted samples.

If two methods pertaining to measure the same quantity are linearly related they may be assumed to measure the same quantity irrespective of the numerical results. This makes recalibration using a reference procedure possible.

10.3 Residuals

Intuitively, the spread of observations is related to the distances between the observations and the estimated regression line. In OLR the relevant distances $(d_{y,x})$ are $y_i - \hat{y}_i$ where $\hat{y}_i$ symbolizes the estimated value from the x_i, i.e., y_i, and the regression line.

These distances are the "residuals" and estimating the standard deviation $(s_{y,x})$ of the residuals follows the same theory as the standard deviation (Eq. 18). Accordingly, the df is $(n - 2)$:

$$
\begin{aligned}
s_{y,x} &= \sqrt{\frac{\sum_{i=1}^{n} (y_i - \hat{y}_i)^2}{n-2}} = \sqrt{\frac{\sum_{i=1}^{n} d_{y,x}^2}{n-2}} = \sqrt{\frac{1}{n-2} \times (SS_{yy} - b \times SS_{xy})} \\
&= \sqrt{\frac{\sum_{i=1}^{n} y_i^2 - n \times y_i^2 - b_{y/x}^2 \times \left(\sum_{i=1}^{n} x_i^2 - n \times (\bar{x})^2 \right)}{n-2}} \\
&= \sqrt{\frac{(n-1) \times \left(s_y^2 - b_{y/x}^2 \times s_x^2 \right)}{n-2}} \\
&= s(y) \times \sqrt{\frac{n-1}{n-2} \times (1 - r^2)}
\end{aligned}
$$

(147)

where $b_{y/x}$ is the slope, s_y and s_x the standard deviation of the y and x variables, respectively. The $d_{y,x}$, i.e.,$(y_i - \hat{y})$ is also recognized as the "error term."

The squared $s_{y,x}$ equals the mean square of the residual variation. Therefore, another route to $s_{y,x}$ is

$$s_{y,x} = \sqrt{\frac{\left(ss_{yy} - \frac{(ss_{xy})^2}{SS_{xx}}\right)}{(n-2)}} \tag{148}$$

In EXCEL the $s_{y,x}$ is calculated by the function *STEYX(y:s, x:s)*, and also shown in the table created by the *LINEST* function (see below).

The $s_{y,x}$ is recognized under different names in the literature, e.g., residual standard deviation (*rsd*) or (s_{res}), residual standard error (*rse*), standard deviation of the line (*sdl*), standard error of the estimate (*see*), or linear residual standard deviation (*ressd*).

10.4 Linear Calibration, Uncertainty of Slope and Intercept

Many quantitative relations between a signal and a concentration in analytical chemistry are linear and the OLR is frequently used in calibrations. There are many exceptions from linearity in measuring systems, e.g., the relation between signal and concentration in immunoassays is rarely linear in the entire measuring interval.

A calibration is usually performed and displayed with the signal on the Y-axis and the concentration on the X-axis. The concentration of the calibrator is usually known with a small, negligible, or zero, uncertainty and therefore the OLR is a suitable model for calibration if the regression is linear. When the calibration is used to convert a signal to concentration, the reverse to the OLR function is used, i.e., the signal (Y) is entered to the calibration function $X = (Y - a)/(b_{y/x})$. In the measurement a value of the signal is entered and the corresponding concentration calculated.

The sample concentration is said to be traceable to the calibrator via the calibration function.

The uncertainty of the signal is transferred to the estimated concentration and the result will have an increased uncertainty compared to that of the calibrator value. This uncertainty is propagated as the grade of the calibrator travels down the traceability chain.

The uncertainty (standard error) of the slope and intercept include the standard deviation of the residuals $s_{y,x}$ (Eq. 147). The formulas can take many different formats:

$$u(b_{y/x}) = \sqrt{\frac{\sum_{i=1}^{n}(y_i - \hat{y}_i)^2}{\sum_{i=1}^{n}(x_i - \bar{x})^2} \times \frac{1}{(n-2)}} = \frac{s_{y,x}}{\sqrt{\sum_{i=1}^{n}(x_i - \bar{x})^2}}$$
$$= \frac{s_{y,x}}{\sqrt{(n-1) \times s(x)^2}} = \frac{s_{y,x}}{\sqrt{SS_{xx}}} \tag{149}$$

where $\hat{y}_i$ is the value of the dependent variable estimated from the corresponding x_i and the regression function and thus $(y_i - \hat{y}_i)$ is the residual at a particular value of x_i.

NB: The uncertainty that is obtained corresponds to the standard error of the slope. Therefore, the confidence interval of the slope is $\pm z \times u(b)$.

The significance of the slope being different from zero (i.e., horizontal) or undetermined (slope $= \infty$) is obtained by calculating the Student's independent t-value. The standard error of the extreme slope is zero and the calculation of the t-value therefore simplified:

$$t = \frac{b_{y/x} - 0}{\sqrt{u(b)^2 - 0}} = \frac{b_{y/x}}{u(b)} \tag{150}$$

The t-value is evaluated using an ordinary t-table and $df = n - 2$. If the slope $(b_{y/x})$ is not different from zero all x-values give the same y-value, i.e., $Y = a$, the intercept or average of the Y-values.

NB: The correlation of the variables can be significant and the coefficient of variation high even if the slope is not significantly different from zero. If the slope is zero or undetermined, however, there cannot be an association between the variables or quantities.

The uncertainty (standard error) of the intercept includes the uncertainty of the residuals, $s_{y,x}$. The formula also comes in many formats:

$$u(a) = s_{y,x} \times \sqrt{\frac{\sum_{i=1}^{n} x_i^2}{n \times (n-1) \times s(x)^2}} = s_{y,x} \times \sqrt{\frac{\sum_{i=1}^{n} x_i^2}{n \times \sum_{i=1}^{n} (x_i - \bar{x})^2}} = s_{y,x}$$

$$\times \sqrt{\frac{1}{n} \times \frac{\sum_{i=1}^{n} (x_i)^2}{SS_{xx}}}$$

$$\tag{151}$$

If the data are displayed in an EXCEL spreadsheet the OLR can be directly shown in the graph, by adding a "trendline." There are also functions to calculate the slope and intercept for a dataset, e.g., $SLOPE(Y_1:Y_n,X_1:X_n)$ and $INTERCEPT(Y_1:Y_n,X_1:X_n)$, where $(Y_1:Y_n,X_1:X_n)$ defines the dataset.

Clearly, if the "sum of squares" are available as defined in Eqs. (142)–(144) and the number of observations are sufficient all characteristics of the regression can be calculated.

A characteristic of the OLR, or "method of linear least squares" is that the variance (uncertainty) of measurements of the independent quantity (X-value) is assumed to be zero or that the ratio between $s(y)^2$ and $s(x)^2$, referred to as λ_i, discussed below Eq. (152), is large. The OLR is relatively robust and acceptable results may be obtained also if the variance of X-values is >0. Further, the variance of the independent variable (Y) should be homoscedastic, i.e., the same variance/standard deviation within the measuring interval.

The OLR is sensitive to outliers and extreme values will therefore have a major impact on the OLR function (see Section 11.10).

The quantity values should ideally be normally distributed in both dimensions and the OLR is regarded as rather robust also in that respect.

Example: Estimate the ORL of the following data

	1	2	3	4	5	6	7	8	9	10	11	12	x(bar)	s(x)
Method 1	1.24	1.34	1.39	1.41	1.64	1.44	1.48	1.51	1.54	1.54	1.54	1.62	1.47	0.12
Method 2	1.3	1.5	1.7	1.5	1.44	1.47	1.6	1.6	1.8	1.5	1.7	1.9	1.58	0.17

From EXCEL the slope and intercept were 0.7856 and 0.4261, respectively. The regression function can also be displayed on the graph as a "trendline" function (Fig. 8).

Using the formulas presented:

$$SS_{xx} = \sum_{i=1}^{n} (x_i - \bar{x})^2 = (n-1) \times s(x)^2 = 11 \times 0.12^2 = 0.148;$$
$$SS_{yy} = 11 \times 0.17^2 = 0.309;$$

$$SS_{xy} = \sum_{i=1}^{n} (x_i - \bar{x}) \times (y_i - \bar{y}) = 0.116$$

$$b_{y/x} = \frac{SS_{xy}}{SS_{xx}} = 0.786; \quad b_{x/y} = \frac{SS_{xy}}{SS_{yy}} = 0.376;$$

$$a = \bar{y} - b \times \bar{x} = 1.58 - 0.785 \times 1.47 = 0.426$$

$$SS_{regr}: \frac{(SS_{xy})^2}{SS_{xx}} = \frac{0.116^2}{0.148} = 0.091: \quad S_{resd} = SS_{tot} - SS_{regr}:$$

$$SS_{yy} - \frac{(SS_{xy})^2}{SS_{xx}} = 0.309 - 0.091 = 0.218$$

The uncertainty of the slope:

$$u(b) = \frac{s_{y,x}}{\sqrt{SS_{xx}}} = \frac{0.148}{\sqrt{0.148}} = 0.383 \; (s_{y,x} \text{ is } 0.148(\text{see Eq. (1.147)}))$$

The uncertainty of the intercept is

$$u(a) = s_{y,x} \times \sqrt{\frac{1}{n} \times \frac{\sum_{i=1}^{n} (x_i)^2}{SS_{xx}}} = 0.148 \times \sqrt{\frac{26.226}{12 \times 0.148}} = 0.567.$$

EXCEL offers the function *LINEST(arrayY;arrayX;const;stat)*. When entered as an array it will give the table below as the output. To carry out an array entry, mark an output area: in the present example, two columns wide (i.e., one column more that the number of independent variables) and five rows and type = and then the LINEST formula beginning with the dependent variable, then the independent variable(s) and then simultaneously press Ctrl, Shift, and Enter. The options *const* and *stat* indicate that the regression line is not forced through the origo (0/0; constant $a = 0$) and if additional statistics should be calculated. The table below will be created.

Slope: 0.786	Intercept: 0.426
SE (slope): 0.383	SE (interc): 0.567
r^2: 0.296	$S_{y,x}$: 0.148
F-value: 4.20	df 10
SS_{regr}: 0.091	SS_{resid}: 0.218

See also under "multivariate regression."

10.5 Deming Regression

In practical work there is usually a variation in the measurements of both the independent and dependent variables. The Deming linear regression (DLR) minimizes the perpendicular (ortho) distances between the observations and a calculated regression line (Fig. 9) by including the ratio between the variance of the independent and dependent observations in the calculation of the formula. If the ratio is equal to 1 then the model minimizes the perpendicular distances from the observations to the regression line. This is the orthogonal regression. The larger the ratio, the more vertical the minimized distance will be and at high ratios it will eventually become vertical and the regression function then becomes identical to the OLR (Fig. 9).

Before the slope (b_D) can be calculated the ratio between the measurement variance of the quantities X and Y, $s(y)^2$ and $s(x)^2$, must be defined

$$\lambda_i = \frac{[s(y)]^2}{[s(x)]^2} \tag{152}$$

Operationally, it is an advantage to also calculate a function (V) that occurs repeatedly in the calculations

$$V = \frac{\sum_{i=1}^{i} (y_i - \bar{y})^2 - \lambda_i \times \sum_{i=1}^{i} (x_i - \bar{x})^2}{2 \times \sum_{i=1}^{i} [(x_i - \bar{x}) \times (y_i - \bar{y})]} = \frac{SS_{yy} - \lambda_i \times SS_{xx}}{2 \times SS_{xy}} \tag{153}$$

$$b_D = V \pm \sqrt{V^2 + \lambda_i} \tag{154}$$

NB: In statistical literature the λ is often defined as $[s(x)]^2/[s(y)]^2$, i.e., $\lambda_i = 1/\lambda$. If that definition is used, then the corresponding changes in Eqs. (153)−(157) shall be made.

The Deming regression approaches the OLR if the $s(y)$ is much larger than $s(x)$ and accordingly $\lambda_i \gg 1$. It may therefore be more convenient to use λ_i as defined in Eq. (150) in these discussions. The Deming regression method, like the OLR, requires that the variance is constant, i.e.,

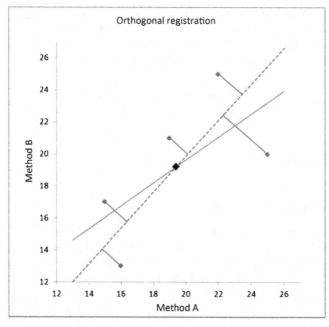

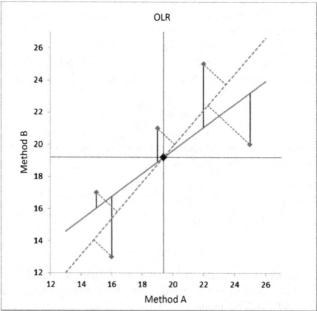

FIGURE 9 Deming (dotted) and Ordinary linear regression (solid) lines. In the left panel the green lines indicate the distances to the regression line that have been minimized. The lambda value (λ_i) is 1 and thus the distance to the regression line is measured perpendicularly to the regression line. The right panel illustrates ($\lambda_i \gg 1$) and accordingly the minimized distances eventually become vertical (*solid blue* (dark gray in print versions)), the residuals larger and the regression identical to the OLR.

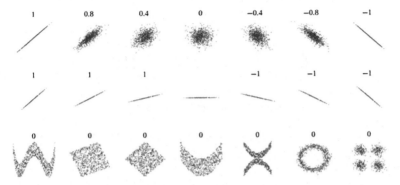

FIGURE 10 The Pearson correlation coefficient calculated for differently distributed datasets. The top row illustrates that the correlation coefficient describes the scatter, but not the slope of a linear regression (middle). The many nonparametric datasets in the bottom row have a correlation coefficient of 0. *From Wikipedia Commons.*

homoscedastic, for both variables and the observations reasonably normally distributed.

Once the slope has been determined the values of the X- and Y averages are inserted and as described for the ORL: $\bar{y} = b_D \times \bar{x} + a$.

The b_D can also be calculated from the ORL regression coefficient Y on X and the Pearson correlation coefficient (r). This is achieved by introducing $b_{y/x}$ and $b_{x/y}$ and rearranging Eq. (153)

$$V = \frac{SS_{yy} - \lambda_i \times SS_{xx}}{2 \times SS_{xy}} = \frac{SS_{yy}}{2 \times SS_{xy}} - \frac{\lambda_i \times SS_{xx}}{2 \times SS_{xy}} = \frac{1}{\dfrac{2 \times SS_{xy}}{SS_{yy}}} - \frac{\lambda_i}{\dfrac{2 \times SS_{xy}}{SS_{xx}}} = \frac{1}{2b_{x/y}} - \frac{\lambda_i}{2b_{y/x}}$$

$$(155)$$

Since

$$\frac{1}{b_{x/y}} = \frac{b_{y/x}}{r^2}; \quad b_{x/y} = \frac{r^2}{b_{y/x}}; \quad r = \sqrt{b_{y/x} \times b_{x/y}}; \tag{156}$$

$$V = \frac{b_{y/x}}{2 \times r^2} - \frac{\lambda_i}{2 \times b_{y/x}} \tag{157}$$

which is then entered into Eq. (155). The intercept is estimated as described in Eq. (140).

Example: Use the same dataset as in the previous section, copied here

	1	2	3	4	5	6	7	8	9	10	11	12	Average	sd
X-value	1.24	1.34	1.39	1.41	1.64	1.44	1.48	1.51	1.54	1.54	1.54	1.62	1.47	0.12
Y-value	1.3	1.5	1.7	1.5	1.44	1.47	1.6	1.6	1.8	1.5	1.7	1.9	1.58	0.17

Assume that $s(x) = s(y)$ in this experiment. Thus, $\lambda_i = 1$.

$$\lambda_i = \frac{[s(y)]^2}{[s(x)]^2} = 1;$$

$$V = \frac{SS_{yy} - \lambda_i \times SS_{xx}}{2 \times SS_{xy}} = \frac{0.309 - 1 \times 0.148}{2 \times 0.116} = 0.692$$

$$b_D = V + \sqrt{V^2 + \lambda_i} = 0.692 + \sqrt{0.692^2 + 1} = 1.90$$

$$a_D = \bar{y} - b_D \times \bar{x} = 1.58 - 1.58 \times 1.47 = -1.29$$

The alternative calculation of the slope is

$$\frac{1}{b_{x/y}} = \frac{b_{y/x}}{r^2}; \quad b_{x/y} = \frac{r^2}{b_{y/x}} = \frac{0.296}{0.786} = 0.377$$

$$V = \frac{b_{y/x}}{2 \times r^2} - \frac{\lambda_i}{2 \times b_{y/x}} = \frac{0.786}{2 \times 0.296} - \frac{1}{2 \times 0.786} = 0.692$$

Etc.

If the λ_i is increased the size of slope will gradually move towards that of the ORL.

When performing these calculations it is essential to retain as many significant digits as possible and only make any rounding in the final result.

There are different formulas to estimate the uncertainty of the slope u_{bD} and intercept u_{aD} and they do not always give the same result.

The uncertainty of the slope of the Deming regression (DLR) is

$$u_{bD} = \sqrt{\frac{b_D^2 \times (1 - r^2)}{r^2 \times (n - 2)}} = \frac{b_D}{r} \times \sqrt{\frac{1 - r^2}{n - 2}} \tag{158}$$

where r is the Pearson correlation coefficient (see Eq. (162))

The uncertainty of the intercept:

$$u_{aD} = \sqrt{\frac{(u_{bD})^2 \times \sum_{i=1}^{n} x_i^2}{n}} \tag{159}$$

In the formulas used here, the r (correlation coefficient) and the sum of the squared results of the independent variable are necessary. They are available in EXCEL: r: $CORREL(Y1:Y12;X1:X12) = 0.543$ and $SUMSQ(X1:X12) = 26.226$, respectively.

$$u_{bD} = \sqrt{\frac{b_D^2 \times (1 - r^2)}{r^2 \times (n - 2)}} = \sqrt{\frac{1.90^2 \times (1 - 0.544^2)}{0.544^2 \times (12 - 2)}} = 1.38$$

$$u_{aD} = \sqrt{\frac{(u_{bD})^2 \times \sum_{i=1}^{n} x_i^2}{n}} = \sqrt{\frac{0.5241 \times 26.23}{12}} = 0.775$$

10.6 Weighted Regression

In a regression analysis it may not be reasonable to assume that every observation shall be treated with equal weights since points far from the centroid (averages of X and Y) have a great influence on the calculations, even if these results are not recognized as outliers. To minimize the influence of outliers and extreme values methods for weighting the data have been developed. The reader is referred to textbooks on statistics for further discussions of weighting.

It is noteworthy that extremes of the independent variable have a larger influence on the regression function, even if the observed dependent variable is within the interval of the remaining observations, than extremes in the dependent variable within the measuring interval of the independent variable.

10.7 Other Regression Functions

10.7.1 Two-Point Calibration

Between two points one, and only one, straight line can be drawn. This is utilized in establishing the common "two point calibration" procedure. This assumes a linear relation between the variables, e.g. signal and concentration. In the calibration procedure, the concentration is measured in two samples, which are well characterized regarding trueness and precision, x_1/y_1 and x_2/y_2:

$$b_{y,x} = \frac{(y_1 - y_2)}{(x_1 - x_2)} = \frac{(y_2 - y_1)}{(x_2 - x_1)} \tag{160}$$

where x_1, x_2 and y_1, y_2 are the corresponding results of the independent and dependent variables, respectively.

The regression (calibration) function is

$$Y - y_1 = b_{y,x} \times (X - x_1) \quad \text{or} \quad Y - y_2 = b_{y,x} \times (X - x_2) \tag{161}$$

Solving any of these equations will give the constant a in $Y = b_{y,x} \times X + a$. If the slope is known the constant can be obtained using the averages of X and Y: $a = b_{y,x} \times \bar{x} - \bar{y}$.

A "two-point calibration" assumes that it is known that the quantities are linearly related.

In routine calibration procedures one of the points is often 0 and although the principles of the two-point calibration is used it may be recognized as a one-point calibration.

10.7.2 Passing-Bablok Regression

Other techniques have been developed to accommodate variances in the results of both variables, and Passing-Bablok regression, similar to the Thiel-Sen estimator, is the most favored. These are nonparametric and thus do not assume any particular distribution of the observed data. Essentially, the Passing-Bablok calculates the slope for all possible lines combining the observations, excluding those which are 0 or indefinite. The slope of the regression line for all observations (*b*) is the median of those of all the connecting lines. The intercept (*a*) is then calculated from the median or average of all observations. As in calculations based on medians the influence of outliers is less than for parametric methods.

The Passing-Bablok regression requires comparatively extensive calculations and is available in some software packages.

10.8 Logistic Regression

There are situations when the effect is dichotomous, i.e., either of two alternatives e.g., "yes" or "no" or 0 and 1, whereas the cause variable is numerical. The similarity with regression analysis is obvious. Suppose the cause variable (independent) is a quantity value and the effect another quantity value but effect above a decision value may be regarded as toxic or not acceptable for other reasons. The difference is really that in OLR we can modify the effect cutoff whereas in problems suitable for logistic regression there are only two outcomes: *yes* or *no*. This will result in a nonlinear relation. It can also be argued, in mathematical terms, that an ORL cannot be used since the right part of the function "$Y = bX + a$" is a real number and as such can take any value whereas the left part can only be one of two results.

This type of problem is frequent in social science and also in medicine but perhaps not in laboratory sciences.

The logistic regression predicts the probability of an event from one or several independent quantities. The probability is expressed as the natural logarithm of the odds. This ratio is known as the logit (Eq. 73). The odds are the ratios of probabilities (*p*) of Y happening (success) to probabilities $(1-p)$ of Y not happening (i.e., failure).

The logistic regression is formulated:

$$\ln\left(\frac{p}{1-p}\right) = \beta_0 + \beta_1 x_1 + \beta_2 x_2 + \cdots + \beta_n x_n$$

Logistic regression solutions are provided in some statistical packages.

The resolved function can be used to estimate the logit for any value of the independent variable. The logit expression shall then be transformed to probabilities to be easily understood. The odds ratio (OR) (Eq. 74) occurs in the logistic regression report and is equivalent to OR $= e^{\beta_1}$.

A probability can only take values between 0 and 1 whereas odds can take values between 0 and infinity. By moving to odds the upper limit of the dependent variable is eliminated and by taking the logarithms the lower limit is eliminated. The logistic regression then becomes a linear relation and can easily be shown in a graph in its simplest form. Higher order regression cannot easily be graphed.

The logistic regression can be used on all types of data.

10.9 Higher Order Regressions

It is not uncommon that a relation between quantities is nonlinear, i.e., the function that describes the relation is of a higher order. Because linear regressions are easy to apply, analysts usually try to linearize the functions, e.g., by transformation of the data See 1.2.4.

10.9.1 Curve Fitting

Complex analytical procedures comprising many and often sequential steps do often not show a linear calibration function. Typical of such reactions are those based on receptor/ligand reactions, often recognized as immunoassays or competitive binding assays. The calibration function is largely sigmoid shaped with an increasing or decreasing slope. This requires a "curve-fitting" procedure and a common approach is fitting the data to a 3- or 4-parameter logistic function (3PL, 4PL). Both functions require that the total ligand binding (top) and nonspecific binding (bottom) are used to define the 50 % binding. The difference is that the 4PL includes a slope factor (Hill factor) which allows modification of the slope, i.e., lengthening or shortening the horizontal segments (top and bottom). The Hill 4PL function is

$$Y = b + \frac{a-b}{1 + (X/c)^d}; \quad X = c \times \left(\frac{a-b}{Y-b}\right)^{1/d}$$

where a, the minimum value that can be obtained (i.e., 0 dose); b, the maximum value that can be obtained (i.e., infinite dose); c, the point of inflection (i.e., halfway between a and b); d, Hill's factor of the curve (i.e., related to the steepness of the curve).

Special software is available and recommended for the curve fitting.

Higher order functions may be fitted to data by special procedures which are available in some statistical packages. EXCEL offers five different trendlines to be fitted to table data. There are software packages which have specialized in "curve-fitting" comprising exponential, logarithmic, polynomial, power etc. regressions.

10.10 Multivariate Linear Regression

An observed effect may depend on more than one independent variable, e.g., the blood pressure may be related to age, weight, P-Cholesterol, P-Triglycerides, and many other causes all of which may be unrelated. Linear multivariate regression analysis offers a method to evaluate the combined effect and the effect of each of the independent variables. Calculations are rather cumbersome and EXCEL and major statistics packages provide ample methods for the calculation.

Multivariate (multivariable) regression usually includes many independent variables (predictors) and one dependent. There shall be the same number of observations for all variables. This corresponds to a model which can be written

$$Y = b_1X_1 + b_2X_2 + \cdots + b_nX_n + a$$

where b represents the "slope" of each of the independent variables and a the intercept. The outcome of a linear multivariate analysis cannot be graphed in a typical scattergram.

The output of a test is usually in the form of a table obtained from an experiment with three independent variables X1, X2, and X3 and 12 observations (Table 14).

The first part of the table is an ANOVA table where regression corresponds to between groups and residual to within group variances. The F test statistic gauges the ability of the regression equation, containing all independent variables, to predict the dependent variable. A large number indicates that the independent variables contribute to the prediction of the dependent

TABLE 14 A Selection of the output of a multiple regression analysis by EXCEL, generated by the regression function in "Data Analysis"

	df	SS	MS	F	p			
Regression	3	0.113	0.038	8.603	0.007			
Residual	8	0.035	0.004					
Total	11	0.148						
Regression Statistics				**Coefficients**	**SE**	**t Stat**	**p-Value**	
Multiple r	0.874		Intercept	1.718	0.296	5.812	0.000	
r^2	0.763		X1	−0.022	0.156	−0.142	0.891	
Adjusted r^2	0.675		X2	−0.279	0.112	−2.478	0.038	
SE	0.066		X3	−0.158	0.073	−2.168	0.062	
Observations	12							

variable. If the F-ratio is around 1, you may conclude that there is no association between the variables, i.e., all the samples are just randomly distributed.

The regression statistics focus on the correlation; the larger the r, the better is the correlation. However, the more predictors you add to a regression analysis the larger will the r become but it does not necessarily mirror any improved prediction, the analysis can be "overfitted." The adjusted r^2 compensates for the increased number of predictors and is therefore a better gauge of a successful optimization the number of predictors. The adjusted r^2 is

$$r^2_{adj} = \frac{\left(1 - r^2\right) \times (N - 1)}{N - k - 1}$$

where N is the number of "pairs" and k is the number of predictors.

By systematically removing predictors which show little or no significance for the prediction (p-value in the right part of the table) the regression is "adjusted" for those predictors, i.e., no longer included in the analysis.

EXCEL offers two possibilities of multiple regression analysis. The more comprehensive analysis is using the function in the "Data analysis" option and the other is using the LINEST function. The latter allows changing the values of the independent and dependent variables but not changing the number of independent variables (Table 15).

More sophisticated software will offer a "predicted r^2" which is generated by systematically removing data, others highlight predictors which seem to be most important.

11. CORRELATION AND COVARIANCE

Correlation describes and quantifies the strength and direction of a linear relationship between two random variables. It is important to consider the difference and connection between regression and correlation.

TABLE 15 Multiple regression analysis results by EXCEL, LINEST function

	X1	X2	X3	Intercept
Coefficients	−0.158	−0.279	−0.022	1.718
SE	0.073	0.112	0.156	0.296
r^2 and *SE*	0.763	0.066		
F	8.603	8.000		
SS	0.113	0.035		

The same data as in Table 14.

11.1 Correlation Coefficient

Regression summarizes the relation between two variables in a regression function but does not address how well the observations fit the function, i.e., the spread (scatter) of the observed pairs in a two-dimensional space. This is quantitatively described by the *correlation coefficient*. The Pearson Product Moment correlation coefficient (r) is applicable to normally distributed data and can be calculated by several, seemingly different, formulas which can be derived from each other but represent different ways to visualize the correlation coefficient:

$$
\begin{aligned}
r &= \frac{\sum_{i=1}^{i=n} [(x_i - \bar{x}) \times (y_i - \bar{y})]}{\sqrt{\sum_{i=1}^{i=n} (x_i - \bar{x})^2 \times \sum_{i=1}^{i=n} (y_i - \bar{y})^2}} \\
&= \frac{n \times \left(\sum_{i=1}^{i=n} (x_i \times y_i) - \sum_{i=1}^{i=n} x_i \times \sum_{i=1}^{i=n} y_i \right)}{\sqrt{\left(n \times \sum_{i=1}^{i=n} x_i^2 - \sum_{i=1}^{i=n} (x_i)^2 \right) \times \left(n \times \sum_{i=1}^{i=n} y_i^2 - \sum_{i=1}^{i=n} (y_i)^2 \right)}} = \\
&= \frac{1}{n-1} \times \sum_{i=1}^{i=n} \left(\frac{(x_i - \bar{x})}{s(x)} \times \frac{(y_i - \bar{y})}{s(y)} \right) = \frac{SS_{xy}}{\sqrt{SS_{xx} \times SS_{yy}}} \\
&= \frac{SS_{xy}}{SS_{xx}} \times \sqrt{\frac{SS_{xx}}{SS_{yy}}} = b_{y/x} \times \sqrt{\frac{SS_{xx}}{SS_{yy}}} \\
&= \left[b_{y/x} \times \sqrt{\frac{SS_{xx}/SS_{xy}}{SS_{yy}/SS_{xy}}} = b_{y/x} \times \sqrt{\frac{b_{x/y}}{b_{y/x}}} = \sqrt{b_{y/x} \times b_{x/y}} \right] \\
&= b_{y/x} \times \sqrt{\frac{\sum_{i=1}^{n} (x_i - \bar{x})^2}{\sum_{i=1}^{n} (y_i - \bar{y})^2}} = b_{y/x} \times \sqrt{\frac{s(x)^2}{s(y)^2}} = b_{y/x} \times \frac{s(x)}{s(y)}
\end{aligned}
$$

$$(162)$$

SS_{xy}, SS_{xx}, and SS_{yy} are defined in Eqs. (142)–(144).

The formula can also be written in yet another form that avoids calculation of the averages:

$$
r = \frac{\sum_{i=1}^{n} x_i \times y_i - \dfrac{\sum_{i=1}^{n} x_i \times \sum_{i=1}^{n} y_i}{n}}{\sqrt{\left(\sum_{i=1}^{n} x_i^2 - \dfrac{\left(\sum_{i=1}^{n} x_i\right)^2}{n} \right) \times \left(\sum_{i=1}^{n} y_i^2 - \dfrac{\left(\sum_{i=1}^{n} y_i\right)^2}{n} \right)}}
\tag{163}
$$

r can assume values between (-1) and ($+1$); i.e., $-1 \le r \le +1$ or $r \le |1|$.

Calculations of r include the average, the standard deviation, or derivatives thereof and thus require that the data are normally or close to normally distributed.

Example: The dataset from the section on regression is used in this example.

	1	2	3	4	5	6	7	8	9	10	11	12	xbar	s(x)
X-value	1.24	1.34	1.39	1.41	1.64	1.44	1.48	1.51	1.54	1.54	1.54	1.62	1.47	0.12
Y-value	1.3	1.5	1.7	1.5	1.44	1.47	1.6	1.6	1.8	1.5	1.7	1.9	1.58	0.17

The correlation coefficient, r, by EXCEL is 0.544 and the slope $b_{y/x}$h-1 = 0.786. Let us apply the first and last expressions in the chain of the visually different but algebraically equal formulas above (162) which give identical results:

$$r = \frac{\sum_{i=1}^{i=n}[(x_i - \bar{x}) \times (y_i - \bar{y})]}{\sqrt{\sum_{i=1}^{i=n}(x_i-\bar{x})^2 \times \sum_{i=1}^{i=n}(y_i-\bar{y})^2}} = \frac{0.116}{\sqrt{0.148 \times 0.309}} = \frac{0.116}{0.214} = 0.544$$

$$r = b_{y/x} \times \frac{s(x)}{s(y)} = 0.786 \times \frac{0.116}{0.167} = 0.544.$$

The correlation coefficient describes the scatter of the observations which is the reverse of the association between the variables. Thus an $r = 0$ indicates no association and $r = 1$ a perfect direct association whereas $r = -1$ indicates a perfect, but inverse, association.

If the variables are centered and expressed as vectors the $\cos(\varphi)$, where φ is the angle between vectors, will be equal to r.

$$\cos(\varphi) = \frac{\sum_{i=1}^{i=n}[(x_i - \bar{x}) \times (y_i - \bar{y})]}{\sqrt{\sum_{i=1}^{i=n}(x_i-\bar{x})^2 \times \sum_{i=1}^{i=n}(y_i-\bar{y})^2}}$$

which is already identified as one of the definitions of r (Eq. 162).

If the regression function is derived from a two-point calibration then $r = 1$. This would be expected since between two defined points one and only one straight line can be drawn.

Even a high correlation coefficient does not imply a causal relationship between the variables.

11.2 Significance of r

The standard error of the correlation coefficient is

$$se(r) = s(\bar{r}) = \frac{1 - r^2}{\sqrt{n - 2}} \tag{164}$$

where n is the number of individuals (samples, i.e., pairs) in the dataset. Strictly, this should only be used for large samples (>100).

The significance of a correlation between two variables is estimated by Student's t-test:

$$|t| = \frac{r}{\sqrt{(1 - r^2)/(n - 2)}} = r \times \sqrt{\frac{n - 2}{1 - r^2}}; \quad df = n - 2 \quad (165)$$

The t-value is evaluated by the usual t distribution table and thus allows estimation of the statistical significance of r.

NB: High values of t, signaling statistical significance, will be obtained with a large number of observations even if r and thus r^2 are small, indicating a limited explanation by the correlation. The interpretation of the r will vary depending on the context. Thus, a $r = 0.7$ ($r^2 = 0.49$) will be regarded as unsatisfactorily low in comparing two measurement procedures in chemistry or physics whereas it might be very differently appreciated in social or medical correlation studies where confounding factors may be more abundant.

The correlation coefficient is not a robust parameter, it is, for instance, much influenced by outliers and assumes a normal distribution of the data.

11.3 Confidence Interval of r

The confidence interval for (r) is estimated after Fisher's transformation of the r; Z is thus a "corrected" r:

$$Z = \frac{1}{2} \times [\ln(1 + r) - \ln(1 - r)] = \frac{1}{2} \times \ln\left(\frac{1 + r}{1 - r}\right); \quad (166)$$

EXCEL offers a function for direct calculation of Z: $= FISHER(r)$.

$$\text{The standard error of Z is } s\left(\overline{Z}\right) = \pm \frac{1}{\sqrt{n - 3}} \quad (167)$$

Thus the confidence interval will be

$$CI_Z = Z \pm z \times \frac{1}{\sqrt{n - 3}} = Z \pm z \times s\left(\overline{Z}\right) \quad (168)$$

where z corresponds to the confidence level, e.g., 1.96 for a 95 % confidence level.

NB: The difference between capital Z and lower case z!

The calculations are based on first converting r to a Z-value (Fisher transformation, also written z') then calculate the confidence interval around Z ($Z \pm CI_z$) and finally convert the endpoints to r by solving Eq. (166) as shown in Eq. (169).

$$r = \frac{e^{2z} - 1}{e^{2z} + 1} \tag{169}$$

NB: The CI_r is not symmetrical around r but the CI_z is symmetrical around Z.

Example: The correlation coefficient, r, for a linear regression of 35 values was 0.91. Calculate the 95 % confidence interval for r.

Fisher's Z value (Eq. 166) is 1.53.

$se_z = \pm 1.96 \times 0.18 = \pm 0.35$. Thus the CI_Z: 1.18−1.88, corresponding to CI_r: 0.83−0.95 when the endpoints of CI_Z are inserted into Eq. (169).

11.4 Coefficient of Determination

The square of the correlation coefficient, also known as the *coefficient of determination* (r^2) is understood as the fraction of the variation in y_i that is accounted for by a linear fit of x_i to y_i; or differently expressed, is the proportion of variance in common between the two variables or shared with the other variable.

For example, for $r = 0.7$, $r^2 = 0.49$ and thus, 49 % of the variation is explained by the linear fit. The coefficient of determination is the quantity that should be interpreted; the correlation coefficient may give an overestimated impression of the association between the variables.

The relation between the residual standard deviation, $s_{y/x}$, and the correlation coefficient and coefficient of determination can be approximated.

$$s_{y/x} = s(y) \times \sqrt{\frac{n-1}{n-2} \times (1 - r^2)};$$
$$\left(\frac{s_{y,x}}{s(y)}\right)^2 = \frac{n-1}{n-2} \times (1 - r^2) \tag{170}$$

If $n - 1$ approaches $n - 2$ then

$$r^2 = 1 - \left(\frac{s_{y,x}}{s(y)}\right)^2 \tag{171}$$

Consequently, and intuitively, the smaller the ratio between $s_{y,x}$ and $s(y)$ the larger becomes r^2. This relation is important to observe in evaluation of a comparison of results by regression analysis. Additional ways to calculate and understand the nature of a correlation coefficient and the coefficient of determination are discussed in the section on regression.

It should be emphasized that correlation does not imply causation. There are many reasons to be careful drawing conclusions from correlation coefficients; even if the p-value indicates a high degree of probability—or significance—this may not give a clue to the root cause. It is essential also to view the correlation in relation to the regression, e.g., in a scatterplot. As shown above Eq. (165), the significance of r is highly dependent on the number of observations.

11.5 Transformation of Regression Data

Many of the properties of regression and correlation studies are calculated assuming that the data is normally distributed. Normal distributions are assumed in the calculation of the slope and correlation coefficient of linear functions. To explore certain features of data which are not linearly or normally distributed the data may be transformed. It may be considered to transform the independent and/or the dependent—or both—variables. This may strengthen the correlation but it should be understood that the correlation between the transformed data is not necessarily numerically the same as a correlation between the original data.

11.6 Spearman Rank Correlation

The Pearson product-moment correlation, r, assumes Gaussian distributed data. If this is not the case, the Spearman's Rank Correlation is used to test the direction and strength of the relationship between two variables.

11.6.1 Spearman's Rank Correlation (r_s) or ρ (Rho)

The Spearman rank correlation coefficient is a nonparametric correlation coefficient. It only addresses the ranks of independently ranked variables. The correlation coefficient is then calculated from the ranks using any of the formulas present in 162. The result is the Spearman correlation coefficient. The Spearman correlation coefficient can also be calculated from the squared differences between the ranks (Eq. 173) if there are no tied results. The Spearman rank coefficient of correlation cannot be interpreted in the same way as the Pearson correlation coefficient because it is based on the ranks and therefore only evaluated on an ordinal scale.

NB: Applying Spearman correlation to a transformed distribution (e.g., logarithm) does not change the coefficient in relation to the coefficient before the transformation since the ranking of the data necessarily remains the same.

If ties occur in the ranking of datasets these can be resolved in two different ways. Either all the ties are given the same rank and the ranking then continues as if each observation were given a rank or alternatively the ties are given the average of the tied ranks as described above in the section of nonparametric procedures. This has been discussed in detail in the section on nonparametric methods.

In previous versions of EXCEL ties could be resolved by the more complex function:

$$IF(ISNUMBER(CR), ((RANK(CR, CR_1:CR_n, 0) + COUNT(CR_1:CR_n) - RANK(CR, CR_1:CR_n, 1) + 1)/2), ""),$$

(172)

where C denotes $\underline{C}$olumn, R $\underline{R}$ow, R_1 the first observation, and R_n the last observation.

Example: Assume a set of pair-wise observations, calculate the Spearman correlation coefficient:

Obs 1	Obs 2	Rank 1	Rank 2	Diff, d_i	Sq Diff
88	105	4	8	4	16
94	93	9	3	−6	36
83	69	2.5	1	−1.5	2.25
91	91	7	2	−5	25
90	107	6	9	3	9
89	100	5	7	2	4
82	96	1	5	4	16
93	99	8	6	−2	4
83	95	2.5	4	1.5	2.25
102	110	10	10	0	0
					114.5

Calculate the ranks in increasing order, for instance using the EXCEL function: $RANK.AVG(R_i, R_1:R_n, 1)$.

Apply the Pearson product moment correlation ($CORREL(y_1:y_n; x_1:x_n)$) to the ranks. This estimates the r_s to 0.304. If, however, the ranking was by RANK.EQ, the r_s equals 0.328.

If there were no ties, e.g., one of the Obs 1 changed from 83 to 81 the simplified method can be used, i.e., calculate the squared difference between the ranks (d_i) and apply

$$\rho \approx r_s = 1 - \frac{6 \times \sum_{i=1}^{n} d_i^2}{n \times (n^2 - 1)}$$

(173)

$$r_s = 1 - \frac{6 \times 115}{10 \times 99} = 1 - 0.697 = 0.370 \quad \text{(no ties)}$$

Incidentally, the Pearson correlation coefficient of the original observations is 0.535.

The significance of a Spearman correlation coefficient (r_s, $df = n - 2$) can be estimated by $t = \sqrt{\dfrac{df \times r_s}{1 - r_s}}$.

11.7 Covariance

The dependence between two random variables is described by the covariance, which is a measure of how two variables vary together. There is no covariance between independent variables. This is not obvious from the formulas and a "covariance" can thus always be calculated but it may be meaningless.

The covariance can be derived from the calculation of variance (Eq. 22) $s^2 = \dfrac{\sum_{i=1}^{i=n}(x_i - \bar{x})^2}{n-1}$ which can be written as $s^2 = \dfrac{\sum_{i=1}^{i=n}[(x_i - \bar{x}) \times (x_i - \bar{x})]}{n-1}$.

The covariance (sample) between x and y is then

$$\text{Covariance} = \frac{\sum_{i=1}^{i=n}[(x_i - \bar{x}) \times (y_i - \bar{y})]}{n-1} = \frac{1}{n-1} \times SS_{xy} = \frac{r \times \sqrt{SS_{xx} \times SS_{yy}}}{n-1} \tag{174}$$

The covariance can become numerically very large. It is calculated for the sample and population with denominator $n-1$ and n, respectively.

Since SS_{xx} and SS_{yy} are $\sum(x_i - \bar{x})^2$ and $\sum(y_i - \bar{y})^2$, respectively, the relation between correlation and covariance can also be expressed by rearranging (Eq. 174)

$$r = \frac{\text{covariance}}{\sqrt{\text{var}(x) \times \text{var}(y)}} = \frac{n \times \text{covariance}}{s(x) \times s(y)} \tag{175}$$

The correlation coefficient may be understood as derived from the covariance and is sometimes expressed as the "normalized covariance."

The importance of covariance can be shown by an example:

Suppose we have two datasets A and B and know the variances of the datasets and their covariance. Then the variance of A + B is VAR(A + B) = VAR(A) + VAR(B) + 2 × COVAR(A,B).

It should be noted that the covariance in Eqs. (174–175) refers to the covariance between observations (x and y-values) not to the covariance between the slope and intercept of the regression function.

The covariance between the slope (b) and the intercept (a) is

$$\text{cov}(a, b) = -\bar{X} \times u\left(b_{y/x}\right)^2 \tag{176}$$

11.8 Correlation and Covariance Matrix

Correlation and covariance can only be calculated between two variables at a time. However, if there is a number of datasets it may be convenient to calculate the correlation and covariance pair-wise between all of them. The outcome is a correlation/covariance matrix. The EXCEL has innate functions (Fig. 11) that are listed under the Add-in—Data analysis, CORRELATION and COVARIANCE, respectively. These commands directly calculate the

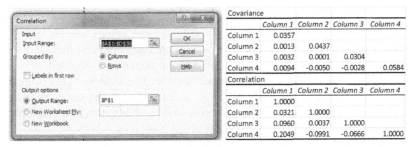

Covariance				
	Column 1	Column 2	Column 3	Column 4
Column 1	0.0357			
Column 2	0.0013	0.0437		
Column 3	0.0032	0.0001	0.0304	
Column 4	0.0094	-0.0050	-0.0028	0.0584
Correlation				
	Column 1	Column 2	Column 3	Column 4
Column 1	1.0000			
Column 2	0.0321	1.0000		
Column 3	0.0960	0.0037	1.0000	
Column 4	0.2049	-0.0991	-0.0666	1.0000

FIGURE 11 Screen dumps of the calculation of the covariance and correlation matrices in EXCEL. The diagonals of the covariance matrix are the variances of the columns whereas the nondiagonals are the covariances. The diagonals of the correlation matrix are 1.

desired matrix comprising the covariance or correlation between all the columns or rows, as specified.

NB: In the covariance matrix the diagonal displays the variances of the data in the columns and the other the covariance terms.

The correlation coefficient and the covariance include the number of observations. Accordingly, EXCEL offers two functions *COVARIANCE.P (interval y_i, interval x_i)* and *COVARIANCE.S(interval y_i, interval x_i)* for calculating the covariance of a population and a sample, respectively. Thus, the population covariance is $(n-1)/n$ times the sample covariance is similar as the difference between sample and population variances. The function in the data analysis operates with the population covariances (Fig. 11).

Many statistics packages also display the various possible scatter diagrams in a similar matrix.

11.9 Outliers

Results of a dataset that appears to differ unreasonably from the rest are called outliers. Outliers are notoriously difficult to handle since theoretically even results far above or below the average or mode may belong to the distribution and should then not be deleted. Identifying outliers therefore comprises a certain subjective judgement. The purpose of the discussion below and the various methods to identify the outliers is to reduce the subjective influence. Outliers may have different root causes and these may be treated differently. If, for instance, the cause is suspected to be a gross error, e.g., erroneous temperature or measuring device, it may be handled differently than if it is suspected to belong to another distribution or that the original distribution was skew. However, if it has an unwarranted effect on the calculated parameters it may be justified to calculate regressions and basic statistics with and without suspected outliers and evaluate the results.

A recommended formal test for outliers is the Grubbs' test in which the statistic G is calculated:

$$G = \frac{\left|\text{suspect value} - \bar{x}\right|}{s} \tag{177}$$

where the average and standard deviation (s) are calculated and including the suspect value. A prerequisite is that the data are normally distributed.

The G is then evaluated using a special table. The formula (178) is applicable if one value is suspected to be an outlier either at the upper or lower end of the distribution. Other formulas are available for more complex situations.

The critical value can be calculated:

$$G > \frac{n-1}{\sqrt{n}} \times \sqrt{\frac{t^2_{(\alpha/(2n),n-2)}}{n-2+t^2_{(\alpha/(2n),n-2)}}} \tag{178}$$

where $t^2_{(\alpha/(2n),n-2)}$ is the critical value of the t-distribution with $n-2$ degrees of freedom and a significance level of $\alpha/2n$. This applies to a two-sided test; for a one-sided test use α/n. If the G is larger than the table value or that calculated from Eq. (178) then the extreme value is unlikely to have occurred by chance. The table can be found in ISO 5725-2.

The Grubbs' test should not be applied if the number of observations is less than 6 or more than 50.

The G-statistic as described above is equal to the z-score (Eq. 59) and can be evaluated as such by using a normal cumulative table or the EXCEL *NORM.S.DIST(G,true)* function.

$$G = \frac{\max|x_i - \bar{x}|}{s(x)}$$

Example: A set of observation had an average of 45.8 and $s(x)$ of 5.1. One observation of 58 was a suspected outlier.

$G = z = \dfrac{58 - 45.8}{5.1} = 2.4$ which is more than the expected $z = 1.96$ for a 97.5 probability of not belonging to the distribution.

NB: The value 58 is above the average and thus prompts a one-sided evaluation. The probability of belonging to the distribution is $(1 - NORM.S.DIST(2.4,TRUE)) = 0.008$ (one-sided).

Another test for outliers is the Dixon test (Q-test) which differs from the Grubbs test by being based on a value being too large (or small) compared to its nearest neighbor:

$$Q = \frac{\left|\text{suspect value} - \text{nearest value}\right|}{\text{range of values}} \tag{179}$$

The critical values can be found in a special table. The Dixon test is usually applicable to $3-10$ observations. The Grubb's and Dixon's tests assume a Gaussian distribution of the quantity values.

Another approach to identifying an outlier is the modified Thompson Tau (τ) test.

First, the absolute value of the deviation (δ) of a suspect result from the average is determined $\delta_i = |x_i - \bar{x}|$, then this is compared with the *tau* value from a table. The tau is calculated as $\tau = \dfrac{t_\alpha \times (n-1))}{\sqrt{n} \times \sqrt{n - 2 + (t_\alpha)^2}}$, where n is the number of data points and $t_{\alpha/2}$ is the critical value of the Student's t-test, e.g., $\alpha = 0.05$ and $df = n - 2$; *(TINV(α,n − 2))*. If $\delta_i > \tau(X)$ the result x_i is an outlier, with the given probability. One sample at a time is considered and the process repeated with a new average, new δ_i and τ, until there are no more outliers suspected or identified.

Outliers in regression analyses or when pairs of dependent data are considered a residual error e_i at the data pair is defined as $e_i = \hat{y} - y$ where $\hat{y}$ is the value of the calculated dependent variable. The statistic that is evaluated is the $e_i/s_{y,x}$, the residual standard deviation. If $\left| e_i/s_{y,x} \right| > 2$ the pair is considered an outlier with a 95 % probability. This metric is regarded as conservative.

The effect of suspect outliers can be minimized by trimming or winsorizing the dataset Eqs. (65) and (66), respectively, but are of less or no importance in nonparametric calculations.

11.10 Leverage

Extreme values and outliers have an influence (*leverage*) on the regression. An observation far from the centroid (average of X and Y) is usually a leverage point but not necessarily an influence point. Influence points have an influence on the regression function and have a tendency to "draw the line closer" but not change the regression function. Leverage points that are not influence points may have a profound effect on the regression, correlation coefficient, and the variance of the variables.

The leverage of an observation x_j can be expressed quantitatively:

$$h_j = \frac{1}{n} + \frac{(x_j - \bar{x})^2}{\sum_{i=1}^{n}(x_i - \bar{x})^2} = \frac{1}{n} + \frac{(x_j - \bar{x})^2}{(n-1) \times s(x)^2} = \frac{1}{n} + \frac{1}{n-1} \times \left(\frac{x_j - \bar{x}}{s(x)}\right)^2$$

$$(180)$$

where $0 \le h_j \le 1$, and n is the number of observations.

The leverage is thus mainly influenced by the distance of the independent variable from the centroid. Leverage is described as a measure of how far away the independent variable values of an observation are from those of the other observations. The last factor in Eq. (180) is equal to the squared z-score Eq. (59) of the observation.

The larger h_j, the more influence it has on the regression. If the observation is in line with a regression line dominated by the bulk of observations (characterized by a small residual) it will have little influence on the regression but on the variance of the distribution of the variables. Often an h_j of 0.9 is used as a cutoff value.

If h_j exceeds

$$h_j > \frac{2 \times p}{n} \quad \text{or} \quad h_j > \frac{3 \times p}{n} \tag{181}$$

where p is the number of predictors (for bivariate linear regression $p = 1$), the point is regarded as a leverage point and needs special consideration.

Extreme high or low quantity values may have a large influence but still not be regarded as "outliers" if they are found on or close to an otherwise defined regression line.

The number of observations in a regression analysis is crucial. The number of observations is directly included in the calculations of the slope and its uncertainty, the intercept, the correlation coefficient and the leverage.

Incidentally, the variance of the predicted value $\hat{y}$ includes part of the leverage:

$$s_{\hat{y}}^2 = s_{y,x}^2 \times \frac{1}{n-1} \times \left(\frac{x_j - \bar{x}}{s_x} \right) \tag{182}$$

12.   COMPARING QUANTITIES

12.1   Comparing Patient Samples

Ideally calibration of different measuring systems with the same calibrator would ascertain the same result if the same quantity were measured in the same sample. For several reasons this is not always the case. Therefore, laboratories or organizations with many instruments compare the performance of measurement procedures using real samples, usually patient samples.

If the same quantities are measured and the results compared, it is fair to assume that the regression will be linear. The calibration itself, however, may take any form and be described by for instance linear, logarithmic, exponential, or polynomial functions. It is thus logical to suspect that if the regression in a comparison is nonlinear and the correlation poor the procedures measure different quantities.

Particularly in clinical experiments the correlation may be poor, often due to imprecision or interfering substances, generally or at certain concentrations. As demonstrated in Eq. (165) high t-values and thus significance may be obtained in a comparison at a given coefficient of variation (r), simply by increasing the number of observations. This prompts for great care before too long-reaching conclusion can be drawn from a significant r-value.

The X-axis represents the equal line in the scatterplot and the regression function is $Y = 0.09X + 0.08$ (see below). The average of the differences (bias) and $\pm 2s$ are shown. The correlation coefficients for the scatter and difference data were 0.748 and 0.019, respectively, illustrating the gain in "resolution" of the differences.

12.2 Difference and Mountain Graphs

It is usually recommended that a scattergram (Fig. 12, left and Fig. 8) is first created in a comparison. This is to give the scientist a broad overview of the distribution of results and to spot possible outliers. Usually the "equal line" and a regression function are also displayed.

To facilitate the evaluation of a comparison, "difference graphs" (Fig. 12, lower panel) are also usually constructed and often demanded for publication in scientific journals. The difference graph displays the difference between the measurements plotted against the average of the results (Bland-Altman) or, if the comparative method (independent variable) can be regarded as a reference procedure, against these values (Var 1 Ind, Fig. 12) directly.

The mechanics behind the design of the difference graph can be understood as subtracting $Y = X$ from the regression $Y = bX + a$, i.e., forming a new function where Y represents the difference and X still represents the comparative method:

$$Y - Y_1 = b \times X - X_1 + a \; ;$$
$$Y = X \times (b - 1) + a \tag{183}$$

As seen from this formula the regression function of the differences will have a slope which is 45 degrees less than that of the original observations. In other words the data of difference graphs appear tilted 45 degrees clockwise in relation to the original regression function. Consequently, the equal line of the original will be represented by the X-axis, the slope of the regression function of the difference graph will be 1 ($TAN(45°)$) less than that of the original data whereas the intercept (a) is unchanged (Fig. 10). The correlation coefficient will be decreased in comparison with that of the original and thus differences appear magnified. There is no unique new information in the difference graph but a visual enhancement. for an example, see legend to Figure 12.

If the differences are normally distributed, the average difference and its standard deviation can be calculated and displayed in the difference graph (Fig. 10) and compared with target values.

Typical questions that are answered by the difference graph, in addition to average and dispersion, are if the difference increases or decreases with the concentration and if the dispersion seems to be constant or change with concentration (Fig. 13).

A complementary graph (Fig. 13), based on the cumulative empirical distribution function for the differences, may be superimposed on the difference graph to better illustrate the distribution of the differences. To accomplish an empirical distribution function, the differences are ranked, any ties resolved,

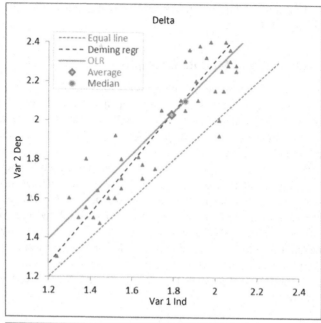

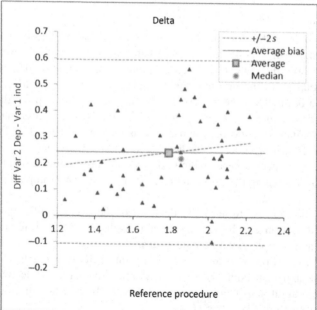

FIGURE 12 Regression and difference graphs. The Deming and OLR functions are shown, $Y = 1.30X - 0.30$ and $Y = 1.09X + 0.08$, respectively. Equal variances ($\lambda_i = 1$) for the methods were assumed for the Deming regression. Both regressions are centered on the average of the dependent and independent variables and therefore the regression lines cross at that point.

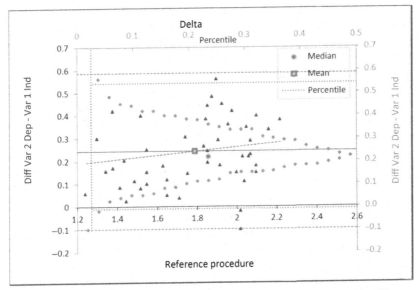

FIGURE 13 The cumulative, empirical distribution function superimposed on the difference graph. The peak of the tilted "mountain" coincides with the median of the differences. The dotted vertical and horizontal line corresponds to the 2.5‰ and 97.5‰, i.e., the central 95% of the observations, centered round the median. Compare Fig. 6!

and the percentiles of the ranks calculated. This produces a sigmoid curve which is an experimental curve and therefore liable to effects of a limited number of observations but approaches the cumulative sigmoid curve of a normal distribution. The function is then mirrored around the 50 % percentile (median) which results in a mountain shaped curve with a sharp peak. Since this is a cumulative curve based on measured "empirical" values, the mountain "peak" corresponds to the median. The mountain can then be tilted 90 degrees clockwise and turned upside down to retain the signs of the difference axis and superimposed on the difference graph. Particularly the tails of the distribution will be easy to evaluate and any skewness revealed.

12.3 Youden Plot

The Youden plot is a graphical method to estimate and visualize random and systematic errors when measuring identical or similar samples of two different concentrations or samples from two or more sites (laboratories or instruments). The original Youden plot was designed to identify the random and systematic errors in many laboratories, e.g., in evaluating EQA or PT (External Quality Assessment, Proficiency Testing) experiments. In these schemes the same samples of two or more concentrations are distributed simultaneously over time, to several participating laboratories.

Generally, the Youden plot is used for evaluating the statistical inference if samples of different concentrations measured in two or more laboratories/instruments. It is also used if duplicate measurements are evaluated although then a normal scatterplot may be more informative. A special feature of the Youden plot are the "help lines." Thus, the graph is subdivided into quadrants by a horizontal and a vertical line through the median of two variables except those recognized as "outliers." This combined median is known as the Manhattan median. A consequence of the design is that there are the same number of observations on both sides of the respective lines.

Another feature is the delineations of acceptance limits. Originally, this was designed as a circle with the radius of 2 standard deviations but nowadays often a rectangle or square is used, which allows different specifications of the measurement procedures their i.e. allowable uncertainty.

The results are plotted (Y_i vs X_i) in a two-dimensional scattergram with the same scales of the axes but adjusted to the concentration levels. Thus, the units become the same on both axes. Sometimes it is an advantage to express the results as z-scores (Eq. 59).

Acceptable results will be found within the nominated area. Results which are located along the positive diagonal but far from the Manhattan median (the coordinates corresponding to the median of the x-values and the y-values) along the positive diagonal indicate a systematic error whereas results far from the diagonal predominantly indicate a random error. However, this distinction is not clear-cut. Results with agreeing results will be found in quadrants I and III, i.e., along the positive diagonal and thus correspond to the "true positive" and "true negative" results. Results in the remaining quadrants represent the "false negative" and the "false positive" results, i.e., error type I and II, respectively.

Concentric circles with the center in the Manhattan median, i.e., the original design, will represent results with identical "total error," defined as. $x^2 + y^2 = r^2$, r representing an unlimited number of combinations of the random and systematic errors. The quadratic addition of two variables results in a loss of information, in particular a negative sign of a variable is lost.

13. PERFORMANCE CHARACTERISTICS

13.1 Definitions

In a dichotomous diagnostic decision situation, e.g., does a person suffer from a given disease or condition or not? there are four possible outcomes. These can be described in a 2×2 frequency table in which the number of individuals belonging to categories (healthy and nonhealthy, respectively) of an independent classification are in rows. The number of individuals or items that are tested positive and negative in each category, respectively, are reported in columns. This layout of the table agrees with the layout of the

TABLE 16 Contingency table, or 2×2 table for classification of dichotomous results assuming a positive test result for disease and negative for nondisease

	Negative Outcome of Test	Positive Outcome of Test	
Disease	False negative (FN)	True positive (TP)	Sensitivity = TP/(TP + FN)
Nondisease	True negative (TN)	False positive (FP)	Specificity = TN/(TN + FP)
	PV(−) = TN/ (TN + FN)	PV(+) = TP/(TP + FP)	Efficiency = (TP + TN)/All

Youden plot and the quadrants. In some literature the categories are in columns and the test in rows. The logics of the present scheme are based on a number line assuming that high test results are concomitant with disease. The design therefore agrees with that of a scatterplot where a high value of the independent variable corresponds to a high value of the independent value and the equal line has a positive slopes (Table 16).

Concordance between diagnostic test results and the independently found diagnosis or property are recorded as True positive = TP, True Negative = TN, False Positive = FP, and False negative = FN. False positives are equal to type 1 error and False negatives to type II errors.

It is of paramount importance in establishing a dichotomous analysis that the diagnoses (categories) are classified independently from the test (columns). A practical example would be plasma cardiac enzyme elevations as the test and post mortem autopsy as the reference and independent end point.

These outcomes are also illustrated in the Youden plot (Fig. 14). The TP will be in the 1st quadrant (I), the FP in the II, the TN in the III, and the FN in the IV quadrant. The True outcomes are thus along the equal line (positive diagonal), i.e., in the odd quadrants, whereas the false outcomes are along the perpendicular diagonal and in the even-numbered quadrants (Fig. 14).

The performance can be expressed as diagnostic (nosographic, clinical) *sensitivity, specificity, predictive value of a positive result* = *PV(+)*, and *predictive value of a negative result* = *PV(−)*. The latter are also called *positive* and *negative predictive values* and abbreviated *PPV* and *PNV*, respectively.

$$\text{Sensitivity(Sens)} = \frac{TP}{TP + FN} \qquad (184)$$

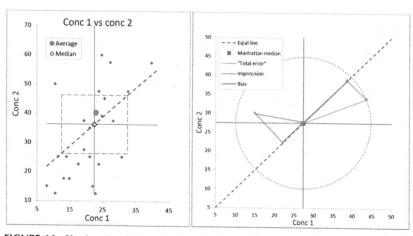

FIGURE 14 Youden plot. Solid vertical and horizontal lines delineate quadrants centered on the median. True positive (Quadrant I, Conc1 high, Conc 2 high), false positive (Quadrant II, Conc 1 low, conc 2 high) true negative (Quadrant III) and false negative (Quadrant IV) are found in the respective quadrants. Left panel: The rectangle illustrates the acceptable deviations, e.g., as uncertainties of the measurement procedures, may be different for the procedures. Right panel: The Manhattan circle and the projection of the "total error" (TE) on the bias and imprecision components. All TE within the circle, by definition, are acceptable. All observations which are localized on the same (concentric) circle have the same TE but with an unlimited number of combinations of systematic (bias) and random (imprecision) errors.

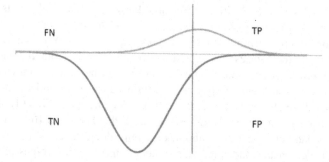

FIGURE 15 Illustration of the contingency table in Table 16. The condition 1 (e.g., a diseased group) is above the X-axis and the condition 2 (e.g., a nondiseased group) below. The vertical line is the "cutoff" of a diagnostic marker. By shifting the cutoff to the right, the specificity increases whereas the sensitivity decreases, provided high values are found in a diseased state. The frequency distributions are idealized, particularly the group of diseased would rather be assumed to be skewed to the right, i.e., a higher frequency of increased results representing various severity of a condition.

$$\text{Specificity(Spec)} = \frac{TN}{TN + FP} \tag{185}$$

The sensitivity (also known as *recall*) and specificity can be changed by changing the cutoff between categories (e.g., considered healthy or non-healthy), reference value, or decision value. In the table this would imply changing the relation between the number of items in the columns of the 2×2 table. An increase in the sensitivity will invariably cause a decrease in the specificity and vice versa. (See also the CDA-plot *Cumulative Distribution Analysis*, Fig. 19).

$$\text{Predictive value of a positive test PV(+)} = \frac{TP}{TP + FP} \qquad (186)$$

$$\text{Predictive value of a negative test PV(-)} = \frac{TN}{TN + FN} \qquad (187)$$

$$\text{Efficiency} = \frac{TP + TN}{TP + TN + FP + FN} = \text{Sens} \times \text{Prev} + \text{Spec} \times (1 - \text{Prev})$$
$$= \text{Prev} \times (\text{Sens} - \text{Spec}) + \text{Spec}$$

$$(188)$$

The *Efficiency* is thus directly proportional to the *Prevalence of disease*. *Efficiency* is also known as "*Index of validity*," "*Agreement*" or *Accuracy*. "*Index of agreement*" is defined as kappa (κ) (Eq. 237)

$$\text{Prevalence of disease (pre} - \text{test probability; Prev)} = \frac{TP + FN}{TP + TN + FP + FN}$$
$$(189)$$

NB: Expressions (186)–(188) depend on the prevalence of disease (Eq. 189) whereas sensitivity (Eq. 184) and specificity (Eq. 185) are characteristics of a specific diagnostic procedure that is used for a specific purpose, using defined discriminators (e.g., cutoffs). Sensitivity and specificity may thus be regarded as constants in that context.

13.1.1 Summary of Terms

The concepts can be interpreted as "rates", i.e., the numerator, the individual outcomes in a contingency table, will be obtained by multiplication with quantities which are often known such as the number of diseased or the number of positive results. Some texts prefer using the quantities expressed as rates.

$$\text{Sensitivity} = \frac{TP}{TP + FN}; \text{ True positive rate (TPR, recall)}$$

$$1 - \text{Sensitivity} = \frac{FN}{TP + FN}; \text{ False negative rate (FNR, miss rate)}$$

$$\text{Specificity} = \frac{TN}{TN + FP}; \text{ True negative rate (TNR)}$$

$$1 - \text{Specificity} = \frac{FP}{TN + FP}; \text{ False positive rate (FPR)}$$

$$1 - PV(+) = \frac{FP}{TP + FP}; \text{ False discovery rate (FDR)}$$

FDR has been developed to rapidly and consistently describe performance of genome analyses.

13.2   Bayes' Theorem

13.2.1   General

The theorem describes a method to estimate the post-test probability by applying key characteristics of an investigation to the pretest probability. In other words: What does experience from similar situations allow us to forecast about the value of a particular test? It is all about probabilities and we need a shorthand or code to describe probabilities. The following rules and definitions may be helpful:

$P(A)$ probability of A. It is a "simple" probability statement, e.g., the probability of rain according to the weather forecast is 60 % would be written $P(rain) = 0.60$.

$P(A|B)$ probability of A if B is known; "Probability of A given B." This is a "conditional" probability, e.g., probability of rain given dark clouds is 65 %, $P(rain|dark\ clouds) = 0.65$.

$P(A \cap B)$ is interpreted "probability A intersection B"; $P(A) \times P(B)$ or probability of A and B. "AND" means multiplication whereas "OR" addition, e.g., "probability of rain **and** thunder is 85 % $P(A \cap B) = 0.85$." *"The probability of rain **or** thunder is 95 %; $P(A \cup B) = 0.95$".*

Since $P(A) \times P(B|A) = P(B) \times P(A|B)$; $\quad P(A|B) = \dfrac{P(A) \times P(B|A)}{P(B)}$

If H is "hypothesis" and E is "evidence" this is translated to

$$P(H|E) = \frac{P(H) \times P(E|H)}{P(E)} \tag{190}$$

13.2.2   This Relationship Justifies a Clarification and Consideration

$P(H|E)$ is the "posterior probability," "hypothesis given evidence," i.e., after receiving more information, i.e., evidence, results of a measurement.

$P(H)$ is the prior probability (in medicine = prevalence) and $P(E|H)$ likelihood. The denominator $P(E)$ is recognized as a normalizing constant. This is the essence of Bayes' theorem, see Eq. (196).

To figure out the probability of a complex event a "probability tree" may be useful.

Example: Suppose three groups of individuals are known for their pipetting skills. Group A consists of 60 % of the individuals, 30 % belong to group B,

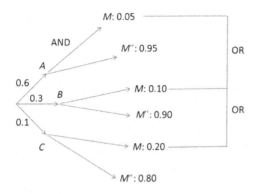

FIGURE 16 Probability tree. Mistakes are *M* and no mistakes are *M″*. Mistakes can occur in group A, B, or C. Numbers refer to conditional probabilities, e.g., probability of no mistakes in group B, $P(M″|B) = 90\ \%$. Oblique arrows (AND) indicate intersection, i.e., $P(A \cap B)$.

and 10 % to group C. It is further known that group A makes 5 % mistakes, group B 10 %, and group C 20 %. A mistake is recognized and the question arises "What is the probability that this was caused in group B?" A mistake is abbreviated *M* and not mistake *M″*, thus $P(M″|B)$?

The known probabilities are summarized in a probability tree (Fig. 16).

In the focus on the left side of the diagram the sum of the probabilities is 1 and they are distributed among the events (groups) A, B, and C, 0.6, 0.3, and 0.1, respectively. In the next split the probabilities of mistakes/not mistakes are shown, e.g., for group A 0.05 and 0.95, respectively.

$$P(M|B) = \frac{P(M) \times P(B|M)}{P(B)} = \frac{0.3 \times 0.1}{0.6 \times 0.05 + 0.3 \times 0.10 + 0.1 \times 0.20} = \frac{0.03}{0.08} = 0.375$$

The probability that the mistake was made in group B was 37.5 %.

The probability for the error occurring in any of the other groups are calculated the similarly and amount to 37.8 % and 25 % for groups A and C, respectively, thus in total 100 %.

13.2.3 The Diagnostic Setting

$$\text{Likelihood ratio}(+)(\text{LR}(+)) = \frac{\text{Sensitivity}}{1 - \text{Specificity}} \qquad (191)$$

$$\text{Likelihood ratio}(-)(\text{LR}(-)) = \frac{1 - \text{Sensitivity}}{\text{Specificity}} \qquad (192)$$

The LR(+) and LR(−) are also known as Bayes' factors.

$$\text{Odds} = \frac{\text{Probability}}{1 - \text{Probability}} = \frac{P(H)}{1 - P(H)} \qquad (193)$$

$$\text{Probability(Prob)} = \frac{\text{Odds}}{1 + \text{Odds}} \qquad (194)$$

Pre-test probability = Prevalence of disease, see Eq. (189)

$$\text{Pretest odds} = \frac{\text{Prevalence}}{1 - \text{Prevalence}} \tag{195}$$

In words, probability is events among all, e.g., diseased among all (Eq. 189) whereas odds is events in relation to non-events, e.g., TP/FN, i.e., a certain reaction in a group in relation the lack of that reaction in the group (Eq. 193).

$$\text{Post-test odds} = \text{Pretest odds} \times \text{LR}(+) \tag{196}$$

This relation summarizes the Bayes' theorem.

$$\text{Post-test probability}[\text{PV}(+) \ \text{or} \ \text{PV}(-)] = \frac{\text{Post-test odds}}{1 + \text{Post-test odds}} \tag{197}$$

$$\text{Prob of disease if a positivetest (PV(+))}: \frac{\text{Sens} \times \text{Prev}}{\text{Sens} \times \text{Prev} + (1 - \text{Spec}) \times (1 - \text{Prev})} \tag{198}$$

$$\text{Prob of no disease if a neg test (PV(-))}: \frac{\text{Spec} \times (1 - \text{Prev})}{\text{Spec} \times (1 - \text{Prev}) + (1 - \text{Sens}) \times \text{Prev}} \tag{199}$$

NB: The post-test probability for a positive and negative result is directly given in the 2 × 2 table as PV(+)and PV(−), respectively.

The post-test probability can be estimated from the prevalence of disease, sensitivity, and specificity as shown in formulas (198) and (199). This will be the same as using formula (196).

NB: probabilities must be transformed to odds, formula (193) and (194), respectively.

A classic example from the clinic is a diagnostic test which has been characterized as correctly identifying X % of diseased individuals (TP). The question then occurs how many false positives (FP) would be anticipated in a population sample where Y % are nondiseased, for instance in a screening project. This question is equal to the probability that a person with a "positive" test is diseased?

To resolve this question three pieces of information are necessary: Sensitivity, Prevalence, and Specificity. To follow the calculations it is crucial to understand that the "false positive rate" (FP) is equal to (1−specificity) times the total number of nondiseased (TN + FP):

$$1 - \text{Specificity} = 1 - \frac{\text{TN}}{\text{TN} + \text{FP}} = \frac{\text{TN} + \text{FP} - \text{TN}}{\text{TN} + \text{FP}} = \frac{\text{FP}}{\text{TN} + \text{FP}}.$$

Thus the frequency or number of nondiseased times (1−Specificity) = FP.

Example: An often cited example is mammography and breast cancer. Mammography is supposed to have a sensitivity of 80 % and a specificity of

90 %. In a sample population the prevalence of breast cancer was assumed to be 1 %. What is the probability that a positive mammography is diagnostic?

For convenience, let us assume an example with a sample of 1,000 individuals. The situation can be illustrated in a tree-structure and a 2×2 table (Fig. 17). Although 80 % of diseased tested positive and only 10 % of the nondiseased, the large number of nondiseased will result in a large percentage of the sample population testing positive without being diseased. Transferred to a percentage of all positives, the predictive value of a positive test is low (7.5 %) which is commonly overlooked (see Fig. 17). This also illustrates the importance of addressing the assumed pretest probability in characterizing a test and the predictive value of a positive test (PV+) in the specific sample.

In mathematical terms it can be written as

$$P((Ca)|(+)) = \frac{P((+)|Ca) \times P(Ca)}{P(Ca) \times P((+)|NotCa)} = \frac{0.8 \times 0.01}{0.8 \times 0.01 + 0.99 \times 0.1} = \frac{8}{107} = 0.075$$

An on-line calculator for quantities related to Bayesian logics is available at http://araw.mede.uic.edu/cgi-bin/testcalc.pl.

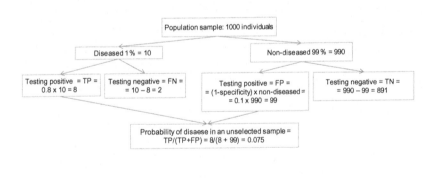

	Negative outcome of test	Positive outcome of test	
Diseased	False negative (2)	True positive (8)	Sensitivity = 0.8
Nondiseased	True negative (891)	False positive (99)	Specificity = 0.90
	PV(−) = 0.998	PV(+) = 0.075	Efficiency = 0.11

FIGURE 17 A tree diagram and a contingency 2×2 table to illustrate the calculation of the probability of positive results in a population given the sensitivity, specificity, and prevalence. The size of the population sample is arbitrary but the prevalence is known (0.01 % or 1 %). Numbers in brackets are estimates from the assumed sample size.

The relation between the pre-test probability and the post-test probability can be visualized in the Fagan nomogram (Fig. 18) in which the likelihood ratio is shown to be the lever between the pre-test probability (prevalence of disease) and the post-test probability and thus indicating the gain in probability by performing the test. The nomogram is based on the relation described in Eq. (196), i.e., $ln(post\text{-}test\ odds) = ln(pretest\ odds) + [ln(LR(+)$ or $LR(-))]$. The odds are recalculated and expressed as $ln(probabilities)$.

A LR(+) of 1 indicates that there is no gain by performing the test. Often an LR(+) of at least 3 is required in clinical work. If the LR(−) is evaluated a small value shall be aimed at.

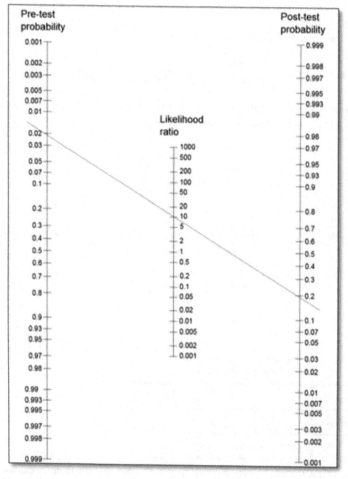

FIGURE 18   Fagan nomogram. Draw a straight line between the pretest probability (2 % in the example) and the LR(+) = 10 and read the post-test probability (20 %) on the right-hand scale.

$$\text{Risk ratio} = \text{relative risk(RR):} \quad RR = \frac{I_e}{I_{ne}} \qquad (200)$$

where I_e is the incidence of exposed individuals and I_{ne} incidence of not exposed individuals. Incidence refers to the number of individuals of a studied group who have contracted a condition (disease) during a specified time. A RR > 1 indicates an increased risk, RR < 1 a benefit of the intervention. RR = 1 then indicates no difference.

13.3 Receiver Operating Characteristics (ROC)

A diagram in which the *diagnostic sensitivity* (*Y*) is plotted against (*1 − diagnostic specificity = False positive rate, FPR*) (*X*) for many chosen different cutoff or reference values is the ROC-curve (Fig. 16).

The ROC-curve is a graphical representation of the trade-off between the true positive and false positive diagnostic rates for every possible quantity value. Equivalently, the ROC-curve is the representation of the trade-offs between sensitivity and specificity. By tradition, the plot shows the false positive rate (*α*) (*1 − specificity*) on the *X*-axis and (1-the false negative rate) i.e., (*1 − β*) (*sensitivity*) on the *Y*-axis.

The ROC-curve also represents the likelihood ratio (LR+) (Eq. 191) for each tested cutoff value.

The sensitivity and specificity concepts as defined above are most suited for binary or dichotomous situations, i.e., a "yes" or "no" answer. This limitation spills over on the ROC-curve. A limitation of the ROC-curve is its negligence of the prevalence of the condition and that cutoff values are not directly displayed. Moving from a continuous response to a dichotomous represents a loss of information, e.g., the severity of a disease.

14. INTERPRETATION OF THE ELEMENTS OF THE ROC ANALYSIS

14.1 Youden Index

If *sensitivity* and *specificity* are diagnostically equally important or desirable, the Youden index [*J*] will indicate the performance (the larger the better) at a given cutoff. Under these circumstances *J* defines an optimal cutoff (*c*).

$$J = \text{Max}_c \left(\text{Sensitivity}_c + \text{specificity}_c - 1 \right) \qquad (201)$$

The maximum value of the *Youden index* is 1 (perfect test) and the minimum is 0 when the test has no diagnostic value. The minimum occurs when *sensitivity = 1 − specificity*, i.e., represented by the equal line (the diagonal) in the ROC diagram. The vertical distance between the equal line and the ROC curve is the *J*-index for that particular cutoff. The *J*-index is represented by the ROC-curve itself.

The optimal outcome is when the *sensitivity* is equal to 1 and at the same time the *specificity* equals 1, i.e., the false positive rate (*1 − specificity*) is zero (0). The point representing this combination will be in the upper left corner of the graph. The closer a *ROC*-curve is to this ideal situation the better the marker performs, given that *sensitivity* and *specificity* are of equal diagnostic importance. This is another way of expressing the Youden index.

Another characteristic is the K-index which is illustrated in the ROC-plot. This index is the distance between the ideal (upper left corner) and the cut.off. Since that is achieved by using the Pythagoras' theorem there are two solutions and the index is illustrated by a quarter-circle. The smaller the K-index the better. Any result above the quarter-circle is superior to the chosen cut-off.

The ROC curve is a summary of information and some information is lost, particularly the actual value of the cutoff. It is therefore important to supplement the ROC curve with the *Cumulative Distribution Analysis (CDA)* (Fig. 19, lower panel) which displays the *sensitivity* and *specificity* against the cutoff values on the X-axis and illustrates the parameters and the cutoff. The CDA is thus a more useful tool in describing the effect of a particular cutoff and its change.

There are additional indices which may be used in the characterization of the diagnostic performance of a test:

Number necessary to test to rule out one diseased (NNT or NNTest).

$$NNT = \frac{TP + TN + FP + FN}{TN}$$

Number necessary to diagnose (NND)

$$NND = \frac{1}{\text{Sensitivity} - (1 - \text{Specificity})} = \frac{1}{\text{Sens} + \text{Spec} - 1} = \frac{1}{\text{Youden index}}$$

These indexes may not be too easy to interpret but obviously, the maximum value of the Youden index is 1, giving NND a value of 1, and the lowest value is zero leading to an undetermined value of NND.

Example: B-Glucose concentrations were measured in 200 patients aged 40−60 years. In this age group the prevalence of disease was estimated to be 5 % by an independent method. The specificity of B-Glucose measurements was estimated to be 0.85 and the sensitivity 0.95. The diagram below (Fig. 20) illustrates the performance if the pretest probability is 5 % (the vertical line) and the effect if the prevalence (pre-test probability) is increased. The *PV(−)* is almost 100 % at this prevalence and is therefore useful for rule-out in a screening situation whereas the predictive value of a positive result is a mere 25 %.

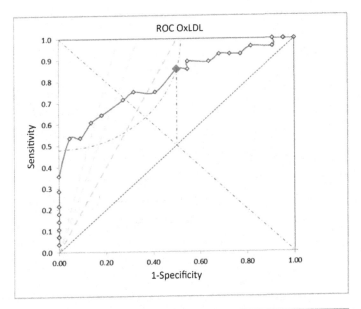

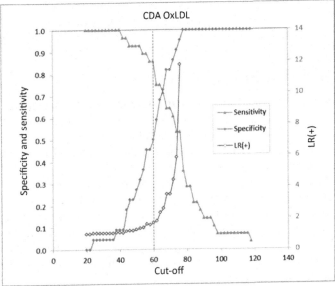

FIGURE 19 ROC- and CDA-curves. In the ROC-curve the quarter-circle represents the K-index and the vertical line the J-index. The *sensitivity* = 1 − *specificity* line represents no diagnostic power and its perpendicular; the theoretical optimum. The diverging hatched lines in the upper panel indicates Bayes' requirements LR(+) = 4, 3, and 2, from left to right. The vertical line in the CDA plot shows the sensitivity, specificity and LR(+) at the chosen cutoff which is not available in the ROC plot.

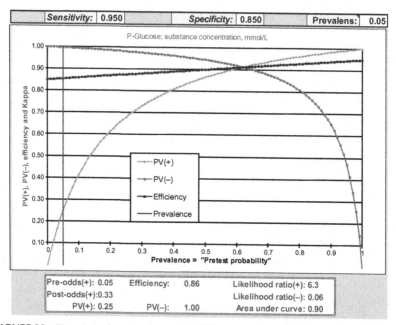

FIGURE 20 The relation between diagnostic performance characteristics and the prevalence of disease. As the vertical line moves horizontally to the right, representing an increasing prevalence of disease, the PV(+) rapidly increases whereas the PV(−) slowly decreases.

14.2   Area Under the Curve (AUC)

The area under the *ROC*-curve (*AUC*) summarizes the diagnostic performance. If the sum of the *sensitivity* and the *specificity* equals one (*TP = FP*) i.e., the *AUC* = 0.5 and the *ROC*-curve follows the diagonal (*J*-index = 0), then the performance is no better than chance. Useful *AUC* are therefore found in the interval 0.5−1.0.

The *AUC* can be estimated by adding the area of trapeziums (a "rectangle with a triangle on top") formed by connecting consecutive points. Visualize a set of columns like in a histogram and estimate their heights! The approximation is in the triangle that remains between the column and the *ROC*-curve. Thus, if (*1 − specificity*) of two adjacent observations is X_1 and X_2 and the corresponding sensitivity Y_1 and Y_2 for the points limiting the trapezium then its area (A_{1-2}) is

$$A_{1-2} = Y_1 \times (X_1 - X_2) + \frac{(Y_2 - Y_1) \times (X_1 - X_2)}{2} = \frac{(X_1 - X_2) \times (Y_1 + Y_2)}{2}$$

(202)

The *AUC* is obtained by adding the individual $A(x_i - x_{(i+1)})$. The more trapeziums that are identified and defined (i.e., the smaller the difference $X_1 - X_2$), the better is the estimate. *AUC* calculation is offered in many software packages. Method performances can be compared by the AUC.

15. OTHER APPLICATIONS

The ROC approach to evaluating signals and responses are not limited to use in medicine but universally applicable. The interpretation of the abbreviation is Receive Operating Characteristics and goes back to the research, development, and description of radar performance before and during WWII.

15.1 Reference Values

Information in a *ROC*-curve can also be used to select a reference value or decision value depending on the priority given to the *sensitivity* or *specificity* in a given clinical setting. Since the theoretical maximal efficiency occurs when *sensitivity* = *specificity* = 1 ($X = 1 -$ Specificity $= 0$; $Y =$ Sensitivity $= 1$, i.e., 0/1), i.e., the upper left corner of the *ROC*-curve, the cutoff corresponding to a minimized distance d (K-index) between the potential reference value and the corner would be an optimal compromise:

$$d = \sqrt{(1-\text{sens})^2 + (1-\text{spec})^2} \qquad (203)$$

In a *ROC*-diagram the d will be represented by a quarter of a circle with its center in 0/1 and the radius d since there are many solutions to Pythagoras' theorem with only d defined. See Fig. 19.

A high *sensitivity* will allow ruling out (*SnNout*) disease if a negative result is obtained (high sensitivity, negative result, out) whereas a high *specificity* rules in (*SpPin*) when the test has a high specificity and a positive result (high specificity, positive result, in). The outcome and thus value of a test is also influenced by the prevalence of disease. See Fig. 20.

Depending on the prevalence, severity, and treatability of the disease it might be more adequate to decide the reference value from a ROC diagram than from the conventional central 95 % interval of a reference population. The choice of a cutoff value may be different depending on the clinical situation. Thus, in a screening situation a high sensitivity may be preferred, if a more specific although more cumbersome method is available for a followup, i.e., the cutoff may be chosen to allow false positive results, i.e., accept a low *specificity* (*SpPin*) rather than compromising *sensitivity* (*SnNout*). In other words, move the cutoff to lower values if high values would signal disease. The rationale may be that it is often better to tell an individual "you may have this disease and we need to make additional investigations" than to dismiss a diseased patient. In the case of a severe but not treatable disease the opposite strategy may be preferred. A reference value based on these criteria may also have economic and political implications.

16. ESTIMATION OF MINIMAL SAMPLE SIZE (POWER ANALYSIS)

16.1 Error Types

Decisions are formally based on formulating a null hypothesis and then testing if this is false or true. Two types of errors can then be made, rejecting the null hypothesis although it is true or accepting the null hypothesis although it is false. There are only two alternatives because it is the null hypothesis we test, it can only be true or false, and the acceptance of the alternative hypothesis will be a consequence. The first effect, also described as detecting an effect that is not, is called a type I error (α), whereas failing to detect an effect is type II (β). The type I and type II errors coincide with the false positive (FP) and false negative (FN) results, respectively., i.e., in quadrants 4 and 2, respectively, of a scatter plot.

The p-value of a significance test equals the probability that a result occurs, i.e., is more extreme than those when the null hypothesis is true. For instance if the null hypothesis is that two values are the same, a high p-value will support that, whereas a small (usually below 0.05) will not and the values are regarded as different with a 95 % probability (two-sided). And there is a 5 % probability ($\alpha = 0.05$) that we make a mistake in this judgment.

Consider two populations in which averages differ by d and which overlap to a certain degree. The null hypothesis assumes that there is no difference between the populations, i.e., $d = 0$. However, a tail of one of the distributions (α) coincides with the other population and a tail of this (β) coincides with the first. Therefore, if one value increases, the other will decrease and the analogy with the diagnostic specificity ($1-\alpha$) and diagnostic sensitivity ($1-\beta$) is obvious (see above).

$$Specificity = 1 - \alpha; \ \alpha = 1 - Specificity \qquad (204)$$

and

$$Sensitivity = power = 1 - \beta; \ \beta = 1 - power \qquad (205)$$

where α is the probability of rejecting a false positive and β is the probability of *not* rejecting a false positive, respectively. That is, α is the false positive rate and β is the false negative rate, usually linked to discussions on the probability of identifying differences (see below).

16.2 Power of a Test

Power is the probability of correctly rejecting the null hypothesis when the alternative is true; with conventional script $P(reject \ H_0|H_1 \ true)$ i.e., a statistical test will reject the null hypothesis when it is false. Therefore, power is equal to $1 - \beta$ or *sensitivity*. A commonly accepted power $= 0.80$, i.e., $\beta = 0.2$.

$$\alpha = \text{``probability of falsely accepting the alternate hypothesis''} \quad (206)$$
$$\text{i.e., rejecting the null hypothesis when true (Type I)}$$

$$\beta = \text{``probability of falsely accepting a null hypothesis''} \quad (207)$$
$$\text{i.e., not rejecting the null hypothesis when it is false (Type II)}$$

$$1 - \beta = \text{``power''} \quad (208)$$

The acceptable size of the ratio β/α is conditional to the purpose of the power analysis; if false positives are critical, increase the ratio, e.g., by decreasing the α, if false negatives are more important, reduce the ratio. An acceptable ratio in clinical practice is often set to $0.20/0.05 = 4$.

16.3 Sample Size

If $\overline{X} - t \times \dfrac{s}{\sqrt{n}} < \mu < \overline{X} + t \times \dfrac{s}{\sqrt{n}}$ then the confidence interval (CI) or margin of error (E) is

$$\text{CI} = t \times \frac{s}{\sqrt{n}} \quad (209)$$

and

$$n = \left(t \times \frac{s}{\text{CI}} \right)^2 \quad (210)$$

The same is valid for proportions (p). If the proportion is not known use $p = 0.5$ that has the highest standard deviation of all proportions.

There are other rules to estimate the sample size. If we want to estimate the sample size from the acceptable standard error of the mean $s(\overline{x})$ this formula can be used:

$$n \approx z_\alpha^2 \times \left(\frac{u(x)}{s(\overline{x})} \right)^2 \quad (211)$$

Example: Suppose the method uncertainty ($u(x)$) is 0.20 and the acceptable standard error of the mean $s(\overline{x}) = 0.5$ then a reasonable number of observations is 31 with 95 % confidence ($\alpha = 0.05$; $z = 2$)

$$n \approx z_\alpha^2 \times \left(\frac{u(x)}{s(\overline{x})} \right)^2 = 1.96 \times \left(\frac{0.20}{0.5} \right)^2 = 31.$$

It is argued to be slightly different to estimate the necessary number of observations to identify a difference between two results (x_1 and x_2).

Example: Assume the same number of observations in the two groups (n) and the same variance $(s(x))^2$ of the results. This leads towards a Student's independent t-test. If the difference between the results of measurements is expressed in standard deviations the expression is

$$d = \frac{x_1 - x_2}{s} \tag{212}$$

The necessary number of observations in each sample is

$$n = \frac{2 \times \left(z_{(1-\alpha/2)} + z_{(1-\beta)}\right)^2}{\left(\frac{\bar{x}_1 - \bar{x}_2}{s(x)}\right)^2} = \frac{2 \times s(x)^2 \times \left(z_{(1-\alpha/2)} + z_{(1-\beta)}\right)^2}{(\bar{x}_1 - \bar{x}_2)^2}$$

$$= \frac{2 \times \left(z_{(1-\alpha/2)} + z_{(1-\beta)}\right)^2}{d^2} \tag{213}$$

If $\alpha = 0.05$ and $\beta = 0.2$, i.e., the $\beta/\alpha = 4$ (see above), $z_{(1-\alpha/2)}$ and $z_{(1-\beta)}$ are 1.96 (in EXCEL: *NORM.S.INV(1 − α/2)*) and 0.84 *(NORM.S.INV (1 − β))*, respectively. The numerator will be 15.7 which is rounded to 16 and a simplified formula (Lehr's quick formula) will be

$$n = \frac{16}{\left(\frac{\bar{x}_1 - \bar{x}_2}{s(x)}\right)^2} = \frac{(s(x))^2 \times 16}{(\bar{x}_1 - \bar{x}_2)^2} = \frac{16}{d^2} \tag{214}$$

where $\bar{x}_1$ and $\bar{x}_2$ are the sample averages, $s(x)$ is the standard deviation of both samples and n the number of observations in each sample. The factor 16 refers to a two-sample situation; in a one-way the factor will be 8.

The detectable difference can be calculated from Eq. (214), if inverted:

$$d^2 = \frac{16}{n}; \quad d = \frac{4}{\sqrt{n}} \tag{215}$$

This is the difference that can be detected with a 95 % confidence for a given sample size.

If there are only two observations the minimum difference will be

$$d \approx \frac{4}{\sqrt{2}} \approx 2.83 = \frac{x_1 - x_2}{s_x}; \quad x_1 - x_2 = 2.83 \times s_x \tag{216}$$

Compare the combined uncertainty of the difference between two observations with the same uncertainty $u(A) = u(C) = z_{1-\alpha/2} \times u(A) \times \sqrt{2} \approx 2.77 \times u(a)$ (cf. Minimum difference (Eq. 90)). This should be exceeded to indicate a significant difference between the results. The factor 4 is an approximation and represents $2 \times (1.96 + 0.84) = 15.68$ which changes the factor in Eq. (216) to 2.77; quod erat demonstrandum.

16.4 Sample Size if Given the % CV

If the % CV is the same for both methods the relative difference is first estimated as

$$RD = \frac{\bar{x}_1 - \bar{x}_2}{(\bar{x}_1 + \bar{x}_2)/2} \tag{217}$$

and the number of samples in each group

$$n \approx \frac{16 \times CV^2}{RD^2} \tag{218}$$

Example: If the relative difference to be detected is, e.g., 20 % and the relative uncertainty (coefficient of variation % CV): 30 %, the following is obtained.

$$n \approx \frac{16 \times \%CV^2}{RD^2} = \frac{16 \times 0.3^2}{\left((1-0.8/1+0.8)/2\right)^2} = \frac{1.44}{0.049} \approx 29$$

If the relevant difference to be detected is 10 % the number of observations needs to be about 130. If the comparison is with a standard, i.e., only one group is needed then the factor in the nominator is 8.

If the averages are known ($\bar{x}_1$ and $\bar{x}_2$) and the % CV specified then

$$n \approx \frac{16 \times (CV)^2}{(\ln(\bar{x}_1) - \ln(\bar{x}_2))^2} \tag{219}$$

16.5 Odds and Risk Ratio

Define
EE and CE, the number of Events in the Experimental and Control groups, respectively,
EN and CN, the number of nonevents in the Experimental and Control groups,
ET and CT, the total number of subjects in the Experimental group and Control groups,
EER and CER, the Experimental Event Rate or probability of the event in the experimental group and the Control Event Rate or probability of the event in the control group;

$$ET = EE + EN; \quad CT = CE + CN; \quad EER = \frac{EN}{ET}; \quad CER = \frac{CN}{CT} \tag{220}$$

Relative risk (Risk ratio cf. Eq.(1.200)):
$$RR = \frac{EER}{CER} = \frac{TP/(FN+TP)}{TN/(TN+FP)} = \frac{Sensitivity}{Specificity} \quad (221)$$

i.e., the ratio between the probability of an event in the experimental and control groups and

$$\text{Experimental event odds: } EEO = \frac{EE}{EN} = \frac{TP}{FN} = \frac{P(pos)}{1 - P(pos)} \quad (222)$$

If the frequency of positive tests among diseased is considered.

$$\text{Control event odds: } CEO = \frac{CE}{CN} \quad (223)$$

The OR is the ratio between the odds for an event in the experimental and control groups.

$$\text{Odds ratio: } OR = \frac{EEO}{CEO} = \frac{EE/EN}{CE/CN} = \frac{EE \times CN}{CE \times EN} = \frac{TP/FN}{TN/FP} = \frac{TP \times FP}{TN \times FN} \quad (224)$$

The relation between OR and RR is

$OR = \frac{RR \times (1 - r_l)}{1 - r_l \times RR}$, where r_l is the larger of the *sensitivity* and *specificity* (Eq. 221).

$$\text{Efficacy of treatment} = 1 - RR \quad (225)$$

$$\text{Number needed to treat or harm: } \frac{1}{EER - CER} \quad (226)$$

if (Eq. 226) < 0 then NNT (Number Needed to Treat);
if (Eq. 226) > 0 then NNH (Number Needed to Harm)

The larger absolute value of the NNT, the less efficient is the treatment.

Example: In a particular study a treatment was compared to no treatment. The outcome summarized in a 2 × 2 table:

	Event (−)	Event (+)	
Treatment	120	20	140
Control	35	110	145
	155	130	

"Sensitivity" $20/140 = 0.143$; "Specificity" $35/145 = 0.241$

$RR = \frac{0.143}{0.241} = 0.593$; $OR = \frac{20 \times 110}{35 \times 120} = 0.524$; $OR = \frac{0.593 \times (1 - 0.241)}{1 - 0.593 \times 0.241} = 0.525$.

17.   AGREEMENT OF CATEGORICAL ASSESSMENTS (KAPPA (κ)-STATISTICS)

This problem is faced when comparing the results of methods on an ordinal scale and the number of agreeing results can be organized in a cross table (contingency table, or frequency table). It is also used when observers are categorizing patients or events into two or several groups, e.g., two or more experts evaluating test results. A special case is the 2×2 frequency table where only two groups are considered (see below).

Any number of categories (groups) can be studied. The table is characterized by the same number of rows and columns (Table 17).

Example: Enter the number of observations in each group for each observer in the appropriate cells:

Proportion of agreement by chance (expected):

$$P_C = \frac{11 \times 21 + 12 \times 22 + 13 \times 23 + 14 \times 24}{N^2} \tag{227}$$

Proportion of observed agreement: $P_O = \dfrac{A + F + K + Q}{N}$ (228)

$$Kappa(\kappa) = \frac{P_O - P_C}{1 - P_C} = 1 - \frac{1 - P_O}{1 - P_C} \tag{229}$$

NB: The *κ-value* is influenced by the *bias* between the observers defined as the difference between Method (Observer) A and Method (Observer) B in their assessment of the frequency of occurrence of a condition. A high value of the *κ-value* indicates a high degree of concordance between observers.

TABLE 17 Notation in the example

		Method (Observer) 1				
		Group 11	Group 12	Group 13	Group 14	Total
Method (Observer) 2	Group 21	A	B	C	D	21
	Group 22	E	F	G	H	22
	Group 23	I	J	K	L	23
	Group 24	M	O	P	Q	24
	Total	11	12	13	14	Total (N)

In a 2×2 table the bias can be quantified as the *Bias Index* (BI), (Symbols in the example are retained, assuming that all input cells except A, B, E, and F are zero [0])

$$BI = \frac{A+B}{N} - \frac{A+E}{N} = \frac{B-E}{N} \qquad (230)$$

thus reflecting the difference between cells of disagreement B and E.

BI results can take values between (-1) and $(+1)$.

Kappa (κ) can be corrected for the bias BAK:

$$BAK = 1 - \frac{1-P_O}{1 - \dfrac{(11+21)^2 + (12+22)^2}{4 \times N^2}} \qquad (231)$$

The value of κ is also affected by the relative probabilities of the "Yes" and "No" answers (in a 2×2 table). This is called *Prevalence Index* (PI):

$$PI = \frac{A-F}{N} \qquad (232)$$

and is the difference between cells of agreement A and F. PI ranges from (-1) to $(+1)$.

A *kappa* (κ) value that is both prevalence and bias adjusted is PABAK:

$$PABAK = 2 \times P_O - 1 \qquad (233)$$

PABAK ranges from (-1) to $(+1)$ like *kappa* (κ) but the interpretation may be different from that of the uncorrected *kappa* (κ) *value*.

The value of κ depends on all these indexes:

$$\kappa = \frac{PABAK - PI^2 + BI^2}{1 - PI^2 + BI^2} \qquad (234)$$

Clearly, if either of BI or PI takes on extreme values, the interpretation of the κ-*value* is difficult.

κ can take any value between -1 and $+1$, $\kappa = 1$ is total agreement, $\kappa = 0$ (agreement expected to chance); $\kappa < 0$ indicates less than expected (rare), total disagreement at $\kappa = (-1)$.

$$se(p) = \pm \sqrt{\frac{P_O \times (1-P_O)}{n \times (1-P_C)^2}} \qquad (235)$$

17.1 Agreement in a 2×2 Table

The efficiency of a diagnostic test is described in a 2×2 table by the *sensitivity* (Eq. 184) and *specificity* (Eq. 185) of the test. The *efficiency* is the sum

TABLE 18 Evaluation of κ-Values

Kappa (κ)	Agreement
0.00	Poor
0.01–0.20	Slight
0.21–0.40	Fair
0.41–0.60	Moderate
0.61–0.80	Substantial
0.81–1.00	Almost perfect

The agreement between test and diagnosis.

of true results relative to all observations (Eq. 188). Considering the influence of chance on the efficiency gives an "expected efficiency":

$$p_e = \frac{(TP + FN) \times (TP + FP) + (TN + FN) \times (TN + FP)}{N^2} \quad (236)$$

Where N is the total number of observations.

The index of agreement i.e., kappa is

$$\kappa = \frac{\text{Efficiency} - p_e}{1 - p_e} \quad (237)$$

κ is interpreted as the difference between the found efficiency and the that expected relative to that possible, considering the chance.

It is noteworthy how chance decreases the efficiency of a test at high and low prevalence of disease. This is particularly important in validating a diagnostic marker, i.e., evaluating it being fit for purpose (Table 18).

The approximate standard error of κ is

$$se(\kappa) = \sqrt{\frac{\text{Efficiency} \times (1 - \text{efficiency})}{n \times (1 - p_e)^2}} \quad (238)$$

Example: Compare two hypothetical 2×2 tables, A and B.

A	Test			B	Test		
	Positive	Negative	Sum		Positive	Negative	Sum
Diseased	10	5	15	Diseased	20	5	25
Healthy	5	80	85	Healthy	5	70	75
Sum	15	85	100	Sum	25	75	100

Sensitivity	0.67	Sensitivity	0.80
Specificity	0.94	Specificity	0.93
Efficiency	0.90	Efficiency	0.90
Expected efficiency	0.75	Expected efficiency	0.63
Kappa (κ)	0.61	Kappa (κ)	0.73

Although the *efficiency* is the same, the κ-values indicate a better agreement between test and diagnosis in the B example which is also demonstrated in the *sensitivity*, i.e., *power*, this time.

Some Metrological Concepts*

METROLOGY, ACCURACY, TRUENESS, AND PRECISION

Quantity

Property of a phenomenon body, or substance, where the property has a magnitude that can be expressed as a number and a reference.

Notes

1. A reference can be a measurement unit, a measurement procedure, a reference material, or a combination of such.
2. The preferred IUPAC-IFCC format for designations of quantities in laboratory medicine is "System—Component; kind-of-quantity."

Kind-of-Quantity

Aspect common to mutually comparable quantities

Note 1 The division of "quantity" according to "klind-of-quantity" is to some extent arbitrary.

Example 1: The quantities diameter, circumference, and wavelength are generally considered to be quantities of the same kind, namely of the kind-of-quantity called length.

Example 2: The quantities heat, kinetic energy, and potential energy are generally considered to be quantities of the same kind, namely of the kind-of-quantiy called energy.

---

* Extracted from JCGM 200:2012 (VIM 2012). *International Vocabulary of Metrology—Basic and General Concepts and Associated Terms (VIM)*, 3rd edition, 2008 version with minor corrections.

This important source of defined terminology is freely downloadable from http://www.bipm.org/utils/common/documents/jcgm/JCGM_200_2012.pdf.

The terms we preset here are not necessarily presented in alphabetical order. The choice of terms is limited and the reader is advised to consult with the original document. Notes and examples may not be cited *in extenso*.

Note 2 Quantities of the same kind within a given system of quantities have the same quantity dimension. However, quantities of the same dimension are not necessarily of the same kind.

Quantity Value

Value of a quantity, value.
Number and reference, together expressing magnitude of a quantity.

Example 1: Length of a given rod: 5.34 m or 534 cm

Example 2: Mass of a given body: 0.152 kg or 152 g

Example 3: Celsius temperature of a given sample: $-5\ ^{\circ}C$

Example 4: Molality of Pb^{2+} in a given sample of water: 1.76 μmol/kg

Example 5: Arbitrary amount-of-substance concentration of lutropin in a given sample of human blood plasma (WHO International Standard 80/552 used as a calibrator): 5.0 IU/L, where "IU" stands for "WHO International Unit"

Note 1 According to the type of reference, a quantity value is—a product of a number and a measurement unit (see Examples 1–4).
The measurement unit one is generally not indicated for quantities of dimension one or

a number and a reference to a measurement procedure or
a number and a reference material (see Example 5).

Measurement

Process of experimentally obtaining one or more quantity values that can reasonably be attributed to a quantity.

Notes

1. Measurement does not apply to nominal properties.
2. Measurement implies comparison of quantities and includes counting of entities.
3. Measurement presupposes a description of the quantity commensurate with the intended use of a measurement result, a measurement procedure, and a calibrated measuring system operating according to the specified measurement procedure, including the measurement conditions.

Measuring System

Set of one or more measuring instruments and often other devices, including any reagent and supply, assembled and adapted to give information used to generate measured quantity values within specified intervals for quantities of specified kinds.

Note

A measuring system may consist of only one measuring instrument.

Measuring Instrument

Device used for making measurements, alone or in conjunction with one or more supplementary devices.

Notes

1. A measuring instrument that can be used alone is a measuring system.
2. A measuring instrument may be an indicating measuring instrument or a material measure.

Measurand

Quantity intended to be measured.

Notes

1. The specification of a measurand requires knowledge of the kind-of-quantity, substance carrying the quantity, including any relevant component, and the chemical entities involved.
2. The measurement, including the measuring system and the conditions under which the measurement is carried out, might change the phenomenon, body, or substance such that the quantity being measured may differ from the measurand as defined.

Example:

1. The length of a steel rod in equilibrium with the ambient Celsius temperature of 23 °C will be different from the length at the specified temperature of 20 °C, which is the measurand. In this case a correction is necessary.
2. In chemistry, "analyte," or the name of a substance or compound, are terms sometimes used for "measurand." This usage is erroneous because these terms do not refer to quantities.

3. The measurand is a *quantity*, e.g., "glucose concentration in serum" where glucose is the *analyte* or *component* and serum is the *system* or *matrix*.

Ordinal Quantity

Quantity, defined by a conventional measurement procedure, for which a total ordering relation can be established, according to magnitude, with other quantities of the same kind, but for which no algebraic operations among those quantities exist.

Example 1: Octane number for petroleum fuel.

Example 2: Subjective level of abdominal pain on a scale from 0 to 5.

   Note 1 Ordinal quantities can enter into empirical relations only and have neither measurement units nor quantity dimensions. Differences and ratios of ordinal quantities have no physical averaging.
   Note 2 Ordinal quantities are arranged according to ordinal quantity-value scales.

Accuracy of Measurement

Measurement accuracy
Accuracy closeness of agreement between a measured quantity value and a true quantity value of a measurand.

Notes

1. The concept "measurement accuracy" is not a quantity and is not given a numerical quantity value. A measurement is said to be more accurate when it offers a smaller measurement error.
2. The term "measurement accuracy" should not be used for measurement trueness and the term measurement precision should not be used for "measurement accuracy," which, however, is related to both these concepts.
3. "Measurement accuracy" is sometimes understood as closeness of agreement between measured quantity values that are being attributed to the measurand.

Trueness

Measurement trueness
Trueness of measurement

Closeness of agreement between the average of an infinite number of replicate measured quantity values and a reference quantity value.

Notes

1. Measurement trueness is not a quantity and thus cannot be expressed numerically, but measures for closeness of agreement are given in ISO 5725.
2. Measurement trueness is inversely related to systematic measurement error, but is not related to random measurement error.
3. Measurement accuracy should not be used for "measurement trueness" and vice versa.

Bias

Measurement bias
Estimate of a systematic measurement error.

Precision

Measurement precision
Closeness of agreement between indications or measured quantity values obtained by replicate measurements on the same or similar objects under specified conditions.

Notes

1. Measurement precision is usually expressed numerically by measures of imprecision, such as standard deviation, variance, or coefficient of variation under the specified conditions of measurement.
2. The "specified conditions" can be, for example, repeatability conditions of measurement, intermediate precision conditions of measurement, or reproducibility conditions of measurement (see ISO 5725-3:1994).
3. Measurement precision is used to define measurement repeatability, intermediate measurement precision, and measurement reproducibility.
4. Sometimes "measurement precision" is erroneously used to average measurement accuracy.

Intermediate Measurement Precision

Intermediate precision
Measurement precision under a set of intermediate precision conditions of measurement.

Intermediate Precision Condition of Measurement

Intermediate precision condition
Condition of measurement, out of a set of conditions that includes the same measurement procedure, same location, and replicate measurements on the same or similar objects over an extended period of time, but may include other conditions involving changes.

Notes

1. The changes can include new calibrations, calibrators, operators, and measuring systems.
2. A specification for the conditions should contain the conditions changed and unchanged, to the extent practical.
3. In chemistry, the term "interserial precision condition of measurement" or "between series imprecision" is sometimes used to designate this concept.

Repeatability

Measurement repeatability
Measurement precision under a set of repeatability conditions of measurement.

Repeatability Condition of Measurement

Repeatability condition
Condition of measurement, out of a set of conditions that includes the same measurement procedure, same operators, same measuring system, same operating conditions, and same location, and replicate measurements on the same or similar objects over a short period of time.

Notes

1. A condition of measurement is a repeatability condition only with respect to a specified set of repeatability conditions.
2. In chemistry, the term "intraserial precision condition of measurement" or "within series imprecision" is sometimes used to designate this concept.

Reproducibility

Measurement reproducibility
Measurement precision under reproducibility conditions of measurement.

Reproducibility Condition of Measurement

Reproducibility condition
Condition of measurement, out of a set of conditions that includes different locations, operators, measuring systems, and replicate measurements on the same or similar objects.

Notes

1. The different measuring systems may use different measurement procedures.
2. A specification should give the conditions changed and unchanged, to the extent practical.

UNCERTAINTY CONCEPT AND UNCERTAINTY BUDGET

Uncertainty

Uncertainty of measurement
Measurement uncertainty
Nonnegative parameter characterizing the dispersion of the quantity values being attributed to a measurand, based on the information used.

Notes

1. Measurement uncertainty includes components arising from systematic effects, such as components associated with corrections and the assigned quantity values of measurement standards, as well as the definitional uncertainty. Sometimes estimated systematic effects are not corrected for but, instead, associated measurement uncertainty components are incorporated.
2. The parameter may be, for example, a standard deviation called standard measurement uncertainty (or a specified multiple of it), or the half-width of an interval, having a stated coverage probability.
3. Measurement uncertainty comprises, in general, many components. Some of these may be evaluated by Type A evaluation of measurement uncertainty from the statistical distribution of the quantity values from series of measurements and can be characterized by standard deviations. The other components, which may be evaluated by Type B evaluation of measurement uncertainty, can also be characterized by standard deviations, evaluated from probability density functions based on experience or other information.

Definitional Uncertainty

Component of measurement uncertainty resulting from the finite amount of detail in the definition of a measurand.

Notes

1. Definitional uncertainty is the practical minimum measurement uncertainty achievable in any measurement of a given measurand.
2. Any change in the descriptive detail leads to another definitional uncertainty.
3. Definitional uncertainty has previously be known as "intrinsic uncertainty."

Uncertainty Budget

Statement of a measurement uncertainty, of the components of that measurement uncertainty, and of their calculation and combination.

Notes

An uncertainty budget should include the measurement model, estimates, and measurement uncertainties associated with the quantities in the measurement model, covariance, type of applied probability density functions, degrees of freedom, type of evaluation of measurement uncertainty, and any coverage factor.

Input Quantity in a Measurement Model

Input quantity
Quantity that must be measured, or a quantity, the value of which can be otherwise obtained, in order to calculate a measured quantity value of a measurand.

Influence Quantity

Quantity that, in a direct measurement, does not affect the quantity that is actually measured, but affects the relation between the indication and the measurement result.

Note

An indirect measurement involves a combination of direct measurements, each of which may be affected by influence quantities.

Example: Amount-of-substance concentration of bilirubin in a direct measurement of hemoglobin amount-of-substance concentration in human blood plasma.

Output Quantity in a Measurement Model

Output quantity
Quantity, the measured value of which is calculated using the values of input quantities in a measurement model.

Expanded Measurement Uncertainty

Expanded uncertainty
Product of a combined standard measurement uncertainty and a factor larger than the number one.

Note

The factor depends upon the type of probability distribution of the output quantity in a measurement model and on the selected coverage probability.

Coverage Interval

Interval containing the set of true quantity values of a measurand with a stated probability, based on the information available.

Notes

1. A coverage interval should not be termed "confidence interval" to avoid confusion with the statistical concept.
2. A coverage interval can be derived from an expanded measurement uncertainty.

Coverage Probability

Probability that the set of true quantity values of a measurand is contained within a specified coverage interval.

Note

The coverage probability is also termed "level of confidence" in the GUM.

Coverage Factor

Number larger than one by which a combined standard measurement uncertainty is multiplied to obtain an expanded measurement uncertainty.

Note

A coverage factor is symbolized k.

MISCELLANEA

Sensitivity

Sensitivity of a measuring system
Quotient of the change in an indication of a measuring system and the corresponding change in a value of a quantity being measured.

Note

1. Sensitivity of a measuring system can depend on the value of the quantity being measured.
2. The change considered in a value of a quantity being measured must be large compared with the resolution.
3. "Analytical sensitivity," i.e., sensitivity of a measuring system should be distinguished from "diagnostic sensitivity" as used in the discussion of Bayes' theorem. "Diagnostic" sensitivity as used in the discussion of Bayes' theorem is not addressed in VIM.
4. "Specificity" is not defined in VIM which defines "selectivity."

Selectivity of a Measuring System

Property of a measuring system, used with a specified measurement procedure, whereby it provides measured quantity values for one or more measurands such that the values of each measurand are independent of other measurands or other quantities in the phenomenon, body, or substance being investigated.

Note

Selectivity used in physics is a concept close to specificity as it is sometimes used in chemistry.

Resolution

Smallest change in a quantity being measured that causes a perceptible change in the corresponding indication.

Note

Resolution can depend on, for example, noise (internal or external) or friction. It may also depend on the value of a quantity being measured.

Detection Limit

Limit of detection
Measured quantity value, obtained by a given measurement procedure, for which the probability of falsely claiming the absence of a component in a material is β, given a probability α of falsely claiming its presence.

Notes

1. IUPAC recommends default values for α and β equal to 0.05.
2. The abbreviation LOD is sometimes used.
3. The term "sensitivity" is discouraged for this concept.

Verification

Provision of objective evidence that a given item fulfils specified requirements.

Examples

1. Confirmation that performance properties or legal requirements of a measuring system are achieved.
2. Confirmation that a target measurement uncertainty can be met.

Notes

1. When applicable, measurement uncertainty should be taken into consideration.
2. The item may be, e.g., a process, measurement procedure, material, compound, or measuring system.
3. Verification should not be confused with calibration. Not any verification is a validation.

Validation

Verification, where the specified requirements are adequate for an intended use.

Example: A measurement procedure, ordinarily used for the measurement of mass concentration of nitrogen in water, may be validated also for measurement in human serum.

Interval

Indication interval
Set of quantity values bounded by extreme possible indications.

Note

An indication interval is usually stated in terms of its smallest and greatest quantity values, e.g., "15−25 mL."

Range of a Nominal Indication Interval

Absolute value of the difference between the extreme quantity values of a nominal indication interval.

Example: For a nominal indication interval of 15−25 mL, the range of the nominal indication interval is 10 mL.
    NB: Colloquially these concepts are often confused. A simple rule is that if two limits are given, it is an "interval", if just one it is a "range." The range does not identify the position of the data on a number line.

Further Reading

There is an abundance of statistical texts related to measurements in laboratories. Some textbooks that have appealed particularly to the author are listed below.

Some of the titles may seem old but their contents is still valid. The reason for including so many is to illustrate the different thinking behind the principles. Several Internet sites provide useful information but are not always reliable. Complete textbooks are also available on the Internet and a selection has been made.

Altman, D. G. *Practical Statistics for Medical Research;* Chapman and Hall: London, 1991–1995. ISBN 978-0-412-27630-9.

Altman, D. G.; Machin, D.; Bryant, T. N.; Gardner, M. J. *Statistics With Confidence*, 2nd ed.; British Medical Journal Publications: London, 2000. ISBN 978-0-7279-1375-3.

Armitage, P.; Berry, G.; Matthews, J. N. S. *Statistical Methods in Medical Research*, 4th ed.; Blackwell Science Ltd: Malden, MA, 2002. ISBN 978-0-6320-5257-8.

Dybkaer, R. *An Ontology of Property for Physical, Chemical and Biological Systems.* http:// ontology.iupac.org.

Engineering Statistics Handbook (NIST 2003). Several updates have been made since. http:// www.itl.nist.gov/div898/handbook/.

Eurachem/CITAC Guide CG4 (QUAM 2000.1). *Quantifying Uncertainty in Analytical Measurement.* http://www.eurachem.org/guides/quam.htm.

Férard, G.; Dybkaer, R.; Fuentes-Arderiu, X. *Compendium of Terminology and Nomenclature of Properties in Clinical Laboratory Sciences. International Federation of Clinical Chemistry and Laboratory Medicine & International Union of Pure and Applied Chemistry;* Silver Book, Royal Society of Chemistry: Cambridge, UK, 2017. ISBN 978-1-78262-107-2.

Field, A. *Discovering Statistics Using IBM SPSS Statistics*, 4th ed.; SAGE Publications: Los Angeles, CA, 2013. ISBN 978-1-4462-4917-8.

Glanz, S. A. *Primer of Biostatistics*, 4th ed.; McGraw Hill: New York, 1996. ISBN 978-0-070-24268-5.

ISO 3534-1:2006. *Statistics—Vocabulary and Symbols. Part 1: General Statistical Terms and Terms Used in Probability*, 2nd ed. *Part 2: Applied Statistics*, 2nd ed.

ISO 5725-2:1994. *Accuracy (Trueness and Precision) of Measurement Methods and Results. Part 2: Basic Method for the Determination of Repeatability and Reproducibility of a Standard Measurement Method.* Geneva 1994, corr 2002.

IUPAC. *IUPAC Compendium of Chemical Terminology—the Gold Book.* https://goldbook.iupac. org/.

Kallner, A. A study of simulated normal probability functions using Microsoft Excel. *Accred. Qual. Assur.* 2016, 21, 271–276.

Kirkwood, B. R.; Sterne, J. A. C. *Essential Medical Statistics*, 2nd ed.; Blackwell Publishing Co: Malden, MA, 2003. ISBN 978-0-865-42871-3.

Lane, D. M. (principal author). *Introduction to Statistics* [Online]. http://onlinestatbook.com/ Online_Statistics_Education.pdf.

Lowry, R. *VassarStat: Concepts & Applications of Inferential Statistics*. http://vassarstats.net/ textbook/.

Miller, J. N.; Miller, J. C. *Statistics and Chemometrics for Analytical Chemistry*, 6th ed.; Pearson Education: Harlow, UK, 2010. ISBN 978-0-273-73042-2.

Taylor, J. R. *An Introduction to Error Analysis: The Study of Uncertainties in Physical Measurements*, 2nd ed.; University Science Books: Sausalito, CA, 1996. ISBN 978-0-935702-75-0.

Van Belle, G. *Statistical Rules of Thumb*, 2nd ed.; John Wiley & Sons, Inc.: Hoboken NJ, 2008. ISBN 978-0-470-14448-0.

Index

Note: Page numbers followed by "*f*" and "*t*" refer to figures and tables, respectively.

Printed in the United States
By Bookmasters